CONTEMPORARY MATHEMATICS

Titles in This Series

Volume

1 **Markov random fields and their applications,** Ross Kindermann and J. Laurie Snell

2 **Proceedings of the conference on integration, topology, and geometry in linear spaces,** William H. Graves, Editor

3 **The closed graph and P-closed graph properties in general topology,** T. R. Hamlett and L. L. Herrington

4 **Problems of elastic stability and vibrations,** Vadim Komkov, Editor

5 **Rational constructions of modules for simple Lie algebras,** George B. Seligman

6 **Umbral calculus and Hopf algebras,** Robert Morris, Editor

7 **Complex contour integral representation of cardinal spline functions,** Walter Schempp

8 **Ordered fields and real algebraic geometry,** D. W. Dubois and T. Recio, Editors

9 **Papers in algebra, analysis and statistics,** R. Lidl, Editor

10 **Operator algebras and K-theory,** Ronald G. Douglas and Claude Schochet, Editors

11 **Plane ellipticity and related problems,** Robert P. Gilbert, Editor

12 **Symposium on algebraic topology in honor of José Adem,** Samuel Gitler, Editor

13 **Algebraists' homage: Papers in ring theory and related topics,** S. A. Amitsur, D. J. Saltman, and G. B. Seligman, Editors

14 **Lectures on Nielsen fixed point theory,** Boju Jiang

15 **Advanced analytic number theory. Part I: Ramification theoretic methods,** Carlos J. Moreno

16 **Complex representations of GL(2, K) for finite fields K,** Ilya Piatetski-Shapiro

17 **Nonlinear partial differential equations,** Joel A. Smoller, Editor

18 **Fixed points and nonexpansive mappings,** Robert C. Sine, Editor

19 **Proceedings of the Northwestern homotopy theory conference,** Haynes R. Miller and Stewart B. Priddy, Editors

20 **Low dimensional topology,** Samuel J. Lomonaco, Jr., Editor

21 **Topological methods in nonlinear functional analysis,** S. P. Singh, S. Thomeier, and B. Watson, Editors

22 **Factorizations of $b^n \pm 1$, $b = 2, 3, 5, 6, 7, 10, 11, 12$ up to high powers,** John Brillhart, D. H. Lehmer, J. L. Selfridge, Bryant Tuckerman, and S. S. Wagstaff, Jr.

23 **Chapter 9 of Ramanujan's second notebook—Infinite series identities, transformations, and evaluations,** Bruce C. Berndt and Padmini T. Joshi

24 **Central extensions, Galois groups, and ideal class groups of number fields,** A. Fröhlich

25 **Value distribution theory and its applications,** Chung-Chun Yang, Editor

26 **Conference in modern analysis and probability,** Richard Beals, Anatole Beck, Alexandra Bellow, and Arshag Hajian, Editors

Titles in This Series

Volume

CONTEMPORARY MATHEMATICS

Complex Differential Geometry and Nonlinear Differential Equations

Proceedings of a Summer Research Conference held August 12–18, 1984

AMERICAN MATHEMATICAL SOCIETY

VOLUME 49

Complex Differential Geometry and Nonlinear Differential Equations

CONTEMPORARY MATHEMATICS

Volume 49

Complex Differential Geometry and Nonlinear Differential Equations

Proceedings of the AMS-IMS-SIAM Joint Summer Research Conference held August 12–18, 1984, with support from the National Science Foundation

Yum-Tong Siu, Editor

AMERICAN MATHEMATICAL SOCIETY

Providence · Rhode Island

The AMS-IMS-SIAM Joint Summer Research Conference in the Mathematical Sciences on Complex Differential Geometry and Nonlinear Differential Equations was held at Bowdoin College, Brunswick, Maine on August 12–18, 1984 with support from the National Science Foundation, Grant DMS-8218075.

1980 *Mathematics Subject Classification*. Primary 53C55; Secondary 32C10, 58G30.

Library of Congress Cataloging-in-Publication Data

AMS-IMS-SIAM Joint Summer Research Conference (1984: Bowdoin College)
Complex differential geometry and nonlinear differential equations.
(Contemporary mathematics, ISSN 0271-4232; v. 49)
Bibliography: p.
1. Global differential geometry—Congresses. 2. Differential equations, Nonlinear—Congresses. I. Siu, Yum-Tong, 1943– . II. American Mathematical Society. III. Institute of Mathematical Statistics. IV. Society for Industrial and Applied Mathematics. V. Title. VI. Series: Contemporary mathematics (American Mathematical Society); v. 49.
QA641.A598 1984 516.3′62 85-28656
ISBN 0-8218-5049-0

This volume was printed directly from author prepared copy.
The paper used in this book is acid-free and falls within the guidelines established to ensure permanence and durability.

LIST OF ARTICLES

PREFACE

These proceedings contain the papers contributed by the speakers of the Summer Research Conference on Complex Differential Geometry and Nonlinear Differential Equations that took place in Bowdoin College, Brunswick, Maine from August 12, 1984 to August 18, 1984. Most of the papers are the actual talks given in the conference. Some papers were developed from the talks and have titles somewhat different from those of the talks to reflect the difference in content. Not every speaker contributed a paper, because some of the talks reported on papers that were already published elsewhere or will soon appear in other publications.

Yum-Tong Siu

Program Schedule

Monday, August 13, 1984

E. Calabi	Yang-Mills problems for Kähler manifolds
J. Spruck	Nonlinear elliptic problems and some geometric applications
Y. Tong	Special harmonic forms on locally Hermitian symmetric spaces
B. White	Harmonic maps and homotopy classes
N. Mok	Kähler-Einstein manifolds of semipositive curvature
P. Wong	Construction of bounded holomorphic functions on simply connected negatively curved Kähler manifolds

Tuesday, August 14, 1984

R. Schoen	Constant mean curvature surfaces in R^3 and S^3
L. Sibner	The removable point singularity problem for Yang-Mills fields
T. Akahori	A new approach to the local embedding theorem of CR structure in the case of $\dim_R M = 2n-1 \geq 7$
R. Sibner	A free boundary minimal surface problem
M. Kalka	The automorphisms of a convex domain
J. Glazebrook	Harmonic maps to quaternionic projective spaces

Wednesday, August 15, 1984

S. Webster	Characteristic numbers of real submanifolds of C^n
N. Stanton	The heat equation for the $\bar{\partial}_b$ equation

Thursday, August 16, 1984

S. Goldberg	Characterizing CP^n by the spectrum of the Laplacian
J. Wolfson	Minimal 2-sphere in complex Grassman manifold
A. Friedman	Blow up of solutions of semilinear parabolic equation
P. Aviles	Conformal deformations of complete manifolds with negative curvature
M. Micallef	Stable minimal surfaces in Euclidean space

Friday, August 17, 1984

J. Sampson	Applications of harmonic maps to Kähler geometry
J. Wood	Holomorphic and harmonic maps and morphisms
M. Anderson	Compactification of minimal submanifolds in E^n
S. T. Yau	Kähler-Einstein metrics with singularities
J. Dodziuk	Laplacian on manifolds and analogous difference operators for graphs
Y. T. Siu	Estimating the dimensions of cohomology groups by curvature

ORGANIZING COMMITTEE

Luis A. Caffarelli Richard M. Schoen Yum-Tong Siu (chairman)

LIST OF PARTICIPANTS

Takao A. Akahori	Max-Planck Institute, West Germany
Michael T. Anderson	Califonia Institute of Technology
Paticio Aviles	University of Illinois
John S. Bland	Tulane University
Eugenio Calabi	University of Pennsylvania
Koji Cho	Harvard University
Hyeong I. Choi	Mathematical Sciences Research Institue, Berkeley
Patick R. Coulton	Eastern Illinois University
Jozef Dodziuk	Queen's College
Franco Favilli	Universita di Pisa, Italy
Alan D. Fekete	Harvard University
Avner Friedman	Northwestern University
James F. Glazebrook	Centro de Investigacion y de Estudios Avanzados del Instituto Politecnio Nacional, Mexico
Samuel I. Goldberg	University of Illinois
David A. Hoffman	University of Massachusetts
Morris Kalka	Tulane University
Luc Lemaire	Free University of Bruxelles
Mario J. Micallef	University of Michigan
Naiming Mok	Princeton University
Nirmala Parkash	Harvard University
Enrique Ramirez de Arellano	Centro de Investigacion y de Estudios Avanzados del Instituto Politecnio Nacional, Mexico

Halsey L. Royden	Stanford University
Joseph H. Sampson	Johns Hopkins University
Richard M. Schoen	University of California, Berkeley
Lesley M. Sibner	Polytech Institute of New York
Robert Sibner	City University of New York
Yum-Tong Siu	Harvard University
Joel Spruck	University of Massachusetts
Nancy K. Stanton	University of Notre Dame
Duraiswamy Sundararaman	Centro de Investigacion y de Estudios Avanzados del Instituto Politecnio Nacional, Mexico
Yue Lin L. Tong	Purdue University
Al Vitter	Tulane University
Sidney M. Webster	University of Minnesota
Brian C. White	Stanford University
Jon G. Wolfson	Rice University
Pit-Mann Wong	University of Notre Dame
John C. Wood	University of Leeds
Paul Yang	University of Southern California
Shing-Tung Yau	University of California, San Diego
Stephen Yau	University of Illinois at Chicago

Contemporary Mathematics
Volume 49, 1986

The new approach to the local embedding theorem of CR-structures for $n \geq 4$ (the local solvablity for the operator $\bar{\partial}_b$ in the abstract sense)

Takao Akahori

ABSTRACT. Kuranishi proved that any abstract strongly pseudo convex CR-structure of real dimension ≥ 9 can be locally embeddable. In (1), by introducing the new approach, we improve Kuranishi's result in (2). Namely, we obtain that any abstract strongly pseudo convex CR-structure of real dimension ≥ 7 can be locally embeddable. In this paper, we briefly sketch the approach in (1).

Introduction. Let $(M, {}^{o}T'')$ be an abstract CR-structure of real dimension ≥ 7. This means that M is a differentiable manifold with real dimension ≥ 7 and ${}^{o}T''$ is a subbundle of complexfied tangent bundle CTM satisfying

1) ${}^{o}T'' \cap {}^{o}\bar{T}'' = 0$, $f\text{-}\dim_C(CTM/({}^{o}T''+{}^{o}\bar{T}'')) = 1$

2) $[\Gamma(M, {}^{o}T''), \Gamma(M, {}^{o}T'')] \subset \Gamma(M, {}^{o}T'')$.

This notion is generalization of a real hypersurface in a complex manifold. Namely, let M be a real hypersurface in a complex manifold N. Then we set

$$ {}^{o}T'' = CTM \cap T''N|_{M} . $$

Then, obviously this pair $(M, {}^{o}T'')$ satisfies 1) and 2). Then, naturally the converse problem arises. Namely, an abstract CR-structure can be embedded as a real hypersurface in a complex

Mathematics Subject Classification 32 G.

Supported by Max Planck Institut für Mathematik and Mittag-Leffler Institute.

manifold. We treat the local problem for this. Namely, an abstract CR-structure can be embedded as a real hypersurface in a complex manifold. We study this problem in the strongly pseudo convex case. For this we must explain the notion of strongly pseudo convex in the abstract sense. By the above condition 1), we can pick a supplementary real vector field S over a neighborhood of the reference point p of M satisfying

$$CTM = {}^{o}T'' + {}^{o}\bar{T}'' + CS \quad \text{over a neighborhood of } p$$

(differentiably) .
By using this decomposition, we can define the Levi-form $L(X,Y)$ by

$$[X,\bar{Y}] \equiv \sqrt{-1}\, L(X,Y)S \mod {}^{o}T''+{}^{o}\bar{T}'' , \text{ for } X , Y \in \Gamma(M, {}^{o}T'') .$$

If eigenvalues of this Levi-form are all positive or all negative, then we call $(M, {}^{o}T'')$ strongly pseudo convex(this notion doesn't depend on the choise of the supplementary vector S).
As is well known, this problem was traeted by several authors, namely, Nirenberg, Andreotti-Hill, Kuranishi, Treves. Especially, by Nirenberg, in the case $n=2$, namely the real dimension three case, there is a counter example. And by Kuranishi, in the case $n \geq 5$, namely real dimension ≥ 9, the local embedding theorem holds. And so the cases $n=3$, $n=4$ were left open. In this paper, I would like to sketch my recent result: the case $n=4$ is also OK, i.e., the local embedding theorem holds in the real dimension 7 case. To see our result, we recall Kuranishi's approach. Kuranishi's approach is divided into two parts.

Part 1. For any approximate C^{∞}-embedding f , he shows that for the CR-structure $(M, {}^{o(f)}T'')$, if $\dim_R M = 2n-1 \geq 7$, on $\wedge({}^{o(f)}T'')^{*}$, D^f - Neumann problem can be solved , and if $\dim_R M = 2n-1 \geq 9$, on $\wedge^2({}^{o(f)}T'')^{*}$, D^f - Neumann problem can be solved over a special kind of neighborhood $U_r(f)$ of p_o , where D^f

means the induced operator by $(M, {}^{o(f)}T'')$ and ${}^{o(f)}T'' = \{X : X \in CTM\ ,\ f_* X \in T''C^n\}$, i.e., the induced CR-structure by f.

Part 2. To find f satisfying: $Df=0$, where D means the induced operator by the given CR-structure $(M, {}^{o}T'')$, he used Nash-Moser's process. Namely by introduction, he constructed a sequence of neighborhoods $U_{r_\mu}(f^{(\mu)})$, and a C^∞-embedding $f^{(\mu)}$ of $U_{r_\mu}(f^{(\mu)})$ into C^n as follows.

$$f^{(1)} = f^{(0)} - M_1 D^{(0)*} N^{(0)} Df^{(0)} \quad \text{on } U_{r_o}(f^{(0)})$$

$$\cdots\cdots$$

$$f^{(\mu+1)} = f^{(\mu)} - M_{\mu+1} D^{(\mu)*} N^{(\mu)} Df^{(\mu)} \quad \text{on } U_{r_\mu}(f^{(\mu)}) ,$$

where M_μ is the smoothing function and $N^{(\mu)} = N^{(f^{(\mu)})}$, obtained in Part 1, $D^{(\mu)} = D^{(f^{(\mu)})}$, and $D^{(\mu)*}$ means the adjoint operator of $D^{(\mu)}$. And if $\dim_R M = 2n-1 \geq 9$, this sequence converges and its limit satisfies $Df=0$.

I must explain why he imposed the assumption $\dim_R M \geq 9$. Roughly speaking, to prove the convergence for $f^{(\mu)}$ by Nash-Moser's process is to show that $Df^{(\mu+1)}$ can be estimated by the quadratic of $Df^{(\mu)}$. And in his proof for this, he used

$$D(f^{(\mu)} - M_{\mu+1} D^{(\mu)*} N^{(\mu)} Df^{(\mu)})$$

$$\fallingdotseq D(f^{(\mu)} - D^{(\mu)*} N^{(\mu)} Df^{(\mu)}) \quad \text{(the notation } \fallingdotseq \text{ means "almost equal")}$$

$$= (D^{(\mu)} - D) D^{(\mu)*} N^{(\mu)} Df^{(\mu)} + D^{(\mu)*} D^{(\mu)} N^{(\mu)} Df^{(\mu)}$$

$$= (D^{(\mu)} - D) D^{(\mu)*} N^{(\mu)} Df^{(\mu)} + D^{(\mu)*} N^{(\mu)} D^{(\mu)} Df^{(\mu)}$$

(we note that here he used $D^{(\mu)} N^{(\mu)} = N^{(\mu)} D^{(\mu)}$ in this equality)

$$= (D^{(\mu)} - D) D^{(\mu)*} N^{(\mu)} Df^{(\mu)} + D^{(\mu)*} N^{(\mu)} (D^{(\mu)} - D) Df^{(\mu)} .$$

And $(D^{(\mu)} - D) Df^{(\mu)}$ behaves like the quadratic of $Df^{(\mu)}$. Therefore in his method, the assumption $\dim_R M = 2n-1 \geq 9$ is necessary (for $D^{(\mu)} N^{(\mu)} = N^{(\mu)} D^{(\mu)}$). Needless to say,

$$\star \qquad D^{(\mu)*} D^{(\mu)} N^{(\mu)} Df^{(\mu)} = N^{(\mu)} D^{(\mu)*} D^{(\mu)} Df^{(\mu)}$$

doesn't make sense because of the boundary condition(we recall that at each step the domain U_{r_μ} varies). However if we had $\star$, we could obtain the local embedding theorem for $\dim_R M=2n-1\geq 7$. Therefore in order to bypass this difficulty,it is very natural to try to reduce our problem to so called " D_b-Neumann problem " , where the boundary condition doesn't appear.

1. D-operator . Let $(M,{}^oT'')$ be given abstract CR-structure . Then we can define a differential operator D as follows . For $u\in\Gamma(M,C)$, we set Du of $\Gamma(M,({}^oT'')^*)$ by

$$Du(X) = Xu \quad , \quad X\in \Gamma(M,{}^oT'') \ .$$

Then like the case for usual scalar valued differential forms , we have a differential complex :

$$0\longrightarrow \Gamma(M,C) \xrightarrow{D} \Gamma(M,({}^oT'')^*) \xrightarrow{D} \Gamma(M,\Lambda^2({}^oT'')^*) \longrightarrow \cdots \quad .$$

Our problem is that : for a given abstracr strongly pseudo convex CR-structure , find a C^∞ -local embedding f at the reference point p_o satisfying : $Df = 0$ on a neighborhood of p_o , where $Df = (Df_1,\ldots,Df_n)$, and $f =(f_1,\ldots,f_n)$.

2. Reducing to D_b-problem . We see how to reduce our CR-embedding theorem . For a given abstract strongly pseudo convex CR-structure $(M,{}^oT'')$, we take an approximate C^∞ -embedding f^o satisfying :

$$j^{(k)}(Df^o)(p_o) = 0 \quad ,$$

where $j^{(k)}$ means k-th jets . We modify f^o . Namely we want to solve

$Du = Df^o$ on a suitable neighborhood $U(p_o)$ of p_o

satisfying :

3) the solution u is estimated by Df^o in a certain sense .

If the above is solved , then f^o-u satisfies

4) $D(f^o-u) = 0$ on a neighborhood $U(p_o)$ of p_o

5) by 3) , f^o-u defines a local embedding at p_o .

Hence for our problem ,i.e.,the local embedding problem , it is enough to find out a neighborhood U satisfying : that U is suitable for D - Neumann problem by using the standard L^2-method(i.e.,Kohn's method) . For this we introduce a family of neighborhoods of p_o ,parametrized by C^∞-embeddings . At first , we set the C^∞-embedding f_ℓ^o satisfying

(6) $f_\ell^o(p_o) = 0$

$$f_\ell^o(M) = \{(z,z_n) : z\in C^{n-1} , \operatorname{Im} z_n - k(z,\operatorname{Re} z_n) = 0\} ,$$

where $k(z,\operatorname{Re} z_n)$ is a real valued C^∞-function satisfying

$$k(z,\operatorname{Re} z_n) = \sum_{1\le i,j\le n-1} (\partial^2 k/\partial z_i \partial \bar z_j)(0) z_i \bar z_j + O(z_\rho, \bar z_\rho, \operatorname{Re} z_n) ,$$

where $O(z_\rho, \bar z_\rho, \operatorname{Re} z_n)$ means the higher order term than 3 , where we regard z_ρ , $\bar z_\rho$ as order 1 and $\operatorname{Re} z_n$ order 2 ,

(7) $O(f_\ell^o)(p_o) = 0$,

(8) $O(f_\ell^o) \in \Gamma(M, {}^o\bar T'' \otimes ({}^oT'')^*)$,

(9) $(1/b(f_\ell^o))^{2\ell} Df_\ell^o$ is bounded near p_o (especially we can assume $\lim_{r_o\to 0} r_o^{-2n}|Y_i f_\ell^o| = 0$, $i=1,2,..,n-1$ from the condition (9)) , where

$$b(f_\ell^o) = \sqrt{\sum_i |Y_i^o t_{f_\ell^o}|^2} , \quad Y_i^o = Y_i^{O(f_\ell^o)} ,$$

$$t_{f_\ell^o} = 2\operatorname{Re}\{(1/2i)(f_\ell^o)_n + (f_\ell^o)_n^{\,2}\} ,$$

where $(f_\ell^o)_n$ means the n-th component of f_ℓ^o in C^n , and $O(f_\ell^o)$ means the induced CR-structure by f_ℓ^o .

(Of course the existence of f_ℓ^o satisfying (6) , (7) , (8) and (9) is not trivial . However we admit this in this paper . For the proof , see Section 1.7 in the paper (1) .) Now we set a family of neighborhoods of p_o ,parametrized by C^∞-embeddings .

$$U_g(r) = \{x , x\in M , t_{f_\ell^o+g}(x)<r\} ,$$

for $g\in H'_{(k),U_{r'}(f_\ell^o)}$,

where $H'_{(k),U_{r'}(f_\ell^o)} = \{g : g$ is a C^∞-map of $U_{r'}(f_\ell^o)$ into C^n , $L_{i_1}^o \ldots L_{i_m}^o g \in L^2$, $m\le k\}$, where $L_i^o = W_i(f_\ell^o)$, $\bar{W}_i(f_\ell^o)$, X^o , Y^o , $\bar{Y}^o$, or the 0-th order operator $(1/b(f_\ell^o))$,

where $W_i(f_\ell^o) = Y_i - (Y_i t_{f_\ell^o}/b(f_\ell^o))Y^o$, namely the tangential part of Y_i with respect to $t_{f_\ell^o}$,

$$Y^o = \sum_{\ell=1}^{n-1} (\bar{Y}_\ell t_{f_\ell^o}/b(f_\ell^o))Y_\ell ,$$

$$X^o = 2\sqrt{-1}\, b(f_\ell^o)S + \bar{\gamma}_{f_\ell^o}Y^o - \gamma_{f_\ell^o}\bar{Y}^o ,$$

$$\gamma_{f^o} = 2\mathrm{Re}(S(h\circ f_\ell^o)) , \text{ where } h\circ f_\ell^o = (1/2i)(f_\ell^o)_n + (f_\ell^o)_n^2$$

and

$$t_{f_\ell^o+g} = 2\mathrm{Re}((1/2i)(f_\ell^o+g)_n + (f_\ell^o+g)_n^2) .$$

We note that if $k \geq 2n-1$ and r is chosen sufficiently small , then

$$C_g = \{x : x \in U(r) , Y_i t_{f_\ell^o+g}(x) = 0 , i=1,2,\dots,n-1\}$$

is just

$$C = \{x : x \in U(r) , Y_i t_{f_\ell^o}(x) = 0 , i=1,2,\dots,n-1 \} ,$$

where $U(r) = \{x : x \in M , t_{f_\ell^o}(x) < r\}$ (because of the definition of $H'_{(k),U_{r'}(f_\ell^o)}$) . So the following first order differential operator makes sense over $U(r)-C$.

$$W_i(f_\ell^o+g) = Y_i - (Y_i t_{f_\ell^o+g}/b(f_\ell^o+g))\sum_\ell (\bar{Y}_\ell t_{f_\ell^o+g}/b(f_\ell^o+g))Y_\ell$$

if r is chosen sufficiently small , where

$$b(f_\ell^o+g) = \sqrt{\sum_i |Y_i t_{f_\ell^o+g}|^2} .$$

With these preparations , we can state our proposition .

<u>Proposition 1.</u> If there is a g of $H'_{(k),U_{r'}(f_\ell^o)}$, $k \geq 2n-1$, satisfying : $W_i(f_\ell^o+g)(f_\ell^o+g) = 0$ on (a neighborhood of p_o) - C , $i=1,2,\dots,n-1$, then

$$U_g(r) = \{x : x \in M , t_{f_\ell^o+g}(x) < r\}$$

is suitable for D -Neumann problem by the L^2 - method(if necessary we must choose r sufficiently small) .

<u>Note I.</u> We note that our partial differential equation $W_i(f_\ell^o+g)(f_\ell^o+g) = 0$, $i=1,2,\dots,n-1$ makes sense on a neighborhood of p_o , U_g , which depends on g . That is to say , for

$g \in H'_{(k)}, U_{r'}(f^o_\ell)$, $W_i(f^o_\ell+g)(f^o_\ell+g)$ is well defined on a neighborhood of p_o, U_g (it is not sure that $W_i(f^o_\ell+g)(f^o_\ell+g)$ is well-defined on $U_r(f^o_\ell)$) . And obviously , our equation is non-linear .

Note II. Even if $f^o_\ell+g$ satisfies our partial non-linear partial differential equation , $f^o_\ell+g$ may not satisfy $D(f^o_\ell+g) = 0$.

Note III. Needless to say , $W_i(f^o_\ell+g)$ means the tangential part of Y_i with respect to $t_{f^o_\ell+g}$.

We sketch the proof of Proposition 1 . In following the standard L^2-method,i.e.,the Kohn's approach for $\bar\partial$ -operator , in our case there is one difficulty . Because

$$\text{f-dim}_C(CTM/({}^oT''_b+{}^o\bar T''_b)) = 2 \quad \text{on} \quad M-C_g$$

where $C_g = \{x:\ x\in M,\ Y_i t_{f^o_\ell+g}(x)=0, i=1,2,..,n-1\}$ and ${}^oT''_b = \{X':\ X'\in {}^oT''_b,\ X't_{f^o_\ell+g} = 0\}$. And in treating with bracket $[W_i(f^o_\ell+g),\bar W_j(f^o_\ell+g)]$, $X^o(g)$ - term , $Y^o(g)-\bar Y^o(g)$ - term might appear , where

$$W_i(f^o+g) = Y_i-(Y_i t_{f^o_\ell+g}/b(f^o_\ell+g))Y^o(g) ,$$

$$Y^o(g) = \sum_{i=1}^{n-1}(\bar Y_i t_{f^o_\ell+g}/b(f^o_\ell+g))Y_i ,$$

$$X^o(g) = \sqrt{-1}\ b(f^o_\ell+g)S + \bar\alpha_g Y^o(g) - \alpha_g \bar Y^o(g) ,$$

where α_g is a C^∞ - function on $U_g(r)-C$ defined by

$$\sqrt{-1}\ b(f^o_\ell+g)S(h\circ(f^o_\ell+g)) + \bar\alpha_g Y^o(g)(h\circ(f^o_\ell+g)) - \alpha_g \bar Y^o(g)(h\circ(f^o_\ell+g)) = 0,$$

where $h\circ(f^o_\ell+g) = (1/2i)(f^o_\ell+g)_n + (f^o_\ell+g)_n{}^2$.

For $X^o(g)$ - term , by using the standard argument , namely by using the Levi-form , we can control this term . But for $Y^o(g)-\bar Y^o(g)$ - term , we have no way to control this term(by the standard argument) . And if the assumption in Proposition 1 holds , then we have

Lemma 2(Key Lemma) . Under the assumption in Proposition 1 ,

$$[W_i(f^o_\ell+g),\bar W_j(f^o_\ell+g)]$$
$$= -\ (\sqrt{-1}/b(f^o_\ell+g))C_S(W_i(f^o_\ell+g),W_j(f^o_\ell+g))X^o(g) + \text{a linear combination}$$

of $W_k(f_\ell^o+g)$, $\bar{W}_k(f_\ell^o+g)$, where $C_S(W_i(f_\ell^o+g), W_j(f_\ell^o+g))$ means the Levi-form with respect to the vector bundle decomposition , introduced in Introduction in this paper ,

$$CTM = {}^oT'' + {}^o\bar{T}'' + CS \quad .$$

Proof . Let $CTM(f_\ell^o+g)$ be a vector bundle on $U_g(r)-C$ defined by

$$\{ X' : X' \in CTM \ , \ X'(h\circ(f_\ell^o+g))=0 \ , \ X'\overline{(h\circ(f_\ell^o+g))}=0 \} .$$

Then , obviously

$$\dim_C CTM(f_\ell^o+g) = 2n-3 \ .$$

In fact , in order to see this , it is enough to see that $d(h\circ(f_\ell^o+g))$ and $d\overline{(h\circ(f_\ell^o+g))}$ are independent at every point near p_o except C . For this , we take the orthonormal base $\{Y_i^{f_\ell^o+g}\}_{1\le i\le n-1}$ of ${}^{o(f_\ell^o+g)}T''$, which means the induced CR-structure by f_ℓ^o+g (of course the existence of such a base is not the trivial fact . However , we admit this in this paper . For the proof , see the paper (1)). And set

$$Y^{f_\ell^o+g} = \sum_\alpha (\bar{Y}_\alpha^{f_\ell^o+g} t_{f_\ell^o+g}/b(f_\ell^o+g))Y_\alpha^{f_\ell^o+g} \quad .$$

Then ,

$$Y^{f_\ell^o+g}(h\circ(f_\ell^o+g)) = 0$$

and

$$Y^{f_\ell^o+g}(h\circ(f_\ell^o+g)) = b(f_\ell^o+g) \quad .$$

Since g is an element of $H'_{(k),U_{r'}(f_\ell^o)}$, $k\ge 2n-1$, $b(f_\ell^o+g)$ doesn't vanish except C . Hence we have that $d(h\circ(f_\ell^o+g))$ and $d\overline{(h\circ(f_\ell^o+g))}$ are independent on $U_r(g)-C$. On the other hand , $\{W_i(f_\ell^o+g)\}_{1\le i\le n-1}$, $X^o(g)$ are of $CTM(f_\ell^o+g)$. In fact ,

$$\begin{aligned}\bar{W}_i(f_\ell^o+g)t_{f_\ell^o+g} &= \bar{W}_i(f_\ell^o+g)((h\circ(f_\ell^o+g)) + \overline{(h\circ(f_\ell^o+g))}) \\ &= \bar{W}_i(f_\ell^o+g)(h\circ(f_\ell^o+g))\end{aligned}$$

(because $W_i(f_\ell^o+g)(f_\ell^o+g) = 0$).

Since $t_{f_\ell^o+g}$ is a real valued C^∞ - function ,

$$\bar{W}_i(f_\ell^o+g)t_{f_\ell^o+g} = 0 \ .$$

So

$$\bar{W}_i(f_\ell^o+g)(h\circ(f_\ell^o+g)) = 0 \ .$$

Hence $CTM(f_\ell^o+g)$ is generated by

$$\{ W_i(f_\ell^o+g), \bar{W}_i(f_\ell^o+g), X^o(g) \} ,$$

because the dimension of the space generated by $W_i(f_\ell^o+g)$, $\bar{W}_i(f_\ell^o+g)$, $X^o(g)$ is 2n-3 . Furthermore

$$[W_i(f_\ell^o+g), \bar{W}_j(f_\ell^o+g)]$$

is of $CTM(f_\ell^o+g)$. Hence there are C^∞ -functions c_{ij} , $a_{k,(i,j)}$, $b_{k,(i,j)}$ satisfying

$$\begin{aligned}[W_i(f_\ell^o+g), \bar{W}_j(f_\ell^o+g)] &= c_{ij}(\sqrt{-1}\, b(f_\ell^o+g)S + \bar{\alpha}_g Y^o(g) - \alpha_g \bar{Y}^o(g)) \\ &+ \sum_k a_{k,(i,j)} W_k(f_\ell^o+g) \\ &+ \sum_k b_{k,(i,j)} \bar{W}_k(f_\ell^o+g) .\end{aligned}$$

By comparing S-term with respect to the C^∞ -vector bundle decomposition

$$CTM = {}^oT'' + {}^o\bar{T}'' + CS ,$$

we have

$$C_S(W_i(f_\ell^o+g), W_j(f_\ell^o+g)) = c_{ij}\sqrt{-1}\, b(f_\ell^o+g) .$$

So we have our lemma . Q.E.D.

With Lemma 2 , by following Kohn's argument , we have an L^2 - estimate over $U_g(r)$ for D - operator (actually we have more strong estimate , and for the details see the paper (1)) . Hence our problem is reduced to the non-linear partial differential equation .

3. The existence of a solution of the non linear partial differential equation . Our main theorem is

Main theorem . If we choose $\ell \geq 10000n^2$, then our non-linear partial differential equation

$$W_i(f_\ell^o+g)(f_\ell^o+g) = 0 , \quad i=1,2,\ldots,n-1 ,$$

has a solution .

For the proof see the paper (1) .

REFERENCES

(1) T.Akahori ,The new approach to the local embedding theorem of CR-structures , the local embedding theorem for $n \geq 4$,I (D^f-estimate) , II (the construction of the solution) , Preprint .

(2) M.Kuranishi , Strongly pseudo convex CR structures over small balls, Part III. An embedding theorem , Annals of Mathematics,116(1982),249-330 .

Mittag-Leffler Institute
Auravägen 17
182 62 DJURSHOLM

Rukyu Univ.
Okinawa,Japan

Contemporary Mathematics
Volume 49, 1986

REMARKS ON CURVATURE INTEGRALS AND MINIMAL VARIETIES

Michael T. Anderson[1]

ABSTRACT. This paper deals with relations between curvature integrals and the global behavior of minimal varieties in Euclidean space, in particular the existence and uniqueness of tangent cones at infinity to such varieties. A number of open problems are discussed.

1. INTRODUCTION. One may separate the complete minimal submanifolds M^n of Euclidean space E^N into two broad classes, according to their behavior at infinity. The first class, denoted $\mathcal{T}(I)$, consists of those minimal immersions which possess "tangent cones at infinity." A simple necessary and sufficient condition for the existence of such cones is

$$\lim_{r\to\infty} \frac{\mathrm{vol}(M\cap B(r))}{r^n} < C$$

for some constant C. Under this condition, one may prove the existence of a sequence $\{r_i\}\to\infty$ such that the varieties $M_{r_i} = \frac{1}{r_i}[M\cap B(r_i)]$ converge to a variety M_∞. M_∞ has the structure of a cone $C(\Sigma)$ (a tangent cone at infinity) on a stationary integral (n-1)-varifold Σ in S^{N-1} (see [2] for a proof). A major open question is whether M_∞ is unique in general. Notice that the structure of the variety Σ at infinity is strongly restricted: the fact that Σ is stationary in S^{N-1} implies that most data, prescribed at infinity, cannot be realized in this fashion.

The second class of minimal varieties, denoted $\mathcal{T}(II)$, are those for which the volume ratio above is unbounded. We note that this division corresponds, in special cases, to well known divisions. Namely, in case $M^n \to C^N$ is complex analytic and non-singular, the division corresponds to M^n algebraic or transcendental (via Stoll's theorem [12]). If $M^2\to E^N$ is a minimal surface of finite topological type, then $M \in \mathcal{T}(I)$ if and only if M has finite total curvature.

1980 Mathematics Subject Classification. 49F20, 53C42, 53C65

[1] N.S.F. Math. Sci. Postdoc. Fellow

It is quite clear that from many points of view, the varieties of $\mathcal{T}(I)$ are much simpler than those of $\mathcal{T}(II)$. In this paper, we collect a number of remarks and observations, as well as open problems concerning the division of $\mathcal{T}(I)$ and $\mathcal{T}(II)$. The main theme is to find characterizations of the distinction $\mathcal{T}(I)/\mathcal{T}(II)$ in terms of the value finite/infinite of curvature integrals attached to M. The case of minimal surfaces ($n = 2$) serves as a model for the higher dimensional case. Of course, the major obstacle at this time to a better understanding is the lack of (non-complex) examples of minimal varieties in dimensions greater than two.

I am very grateful to Frank Morgan for pointing out an error in an earlier version of this paper.

2. CURVATURE INTEGRALS. In this section, we discuss briefly a number of curvature integrals that one may attach to complete submanifolds $M^n \to E^N$. A natural class of integrals are those that occur as coefficients in Weyl's formula for the volume of a tube $T_r(\Omega)$ of radius r about $\Omega \subset M$ (see [7],[13]):

$$\text{vol } T_r(\Omega) = \text{vol}\Omega \cdot r^{N-n} + \sum_{j=2}^{n} c_j \int_\Omega \text{Tr}(\Lambda^j R)\cdot r^{N-n+j} , \tag{2.1}$$

where R is the curvature tensor of M and $\Lambda^j R \in \text{Hom}[\Lambda^p TM, \Lambda^p TM]$, $p = j/2$, is the induced operator on p-forms. Note that vol $T_r(\Omega)$ is counted with multiplicity arising from cut points of the normal exponential map.

In order to have scale invariant quantities, let $B(r)$ denote the ball in E^N of radius r about 0; set

$$I_k(r) = \frac{1}{r^{n-2k}} \int_{M\cap B(r)} \text{Tr}(\Lambda^{2k} R) \tag{2.2}$$

and $I_k(M) = \sup_r |I_k(r)|$. Note that $I_0(M) = \sup_r \frac{\text{vol}(M\cap B(r))}{r^n}$ and $I_{n/2} = \sup_r \left|\int_{M\cap B(r)} \Omega\right|$, where Ω is the Chern-Gauss-Bonnet form of M. In case $N = n+1$, the integrand $\text{Tr}(\Lambda^j R)$ is the jth elementary symmetric function of the eigenvalues of the second fundamental form A of $M^n \to E^{n+1}$. For a minimal submanifold of E^N, one has

$$I_1(r) = \frac{1}{r^{n-2}} \int_{M\cap B(r)} (-\tau) = \frac{1}{r^{n-2}} \int_{M\cap B(r)} |A|^2 \tag{2.3}$$

where τ is the scalar curvature of M, appropriately normalized.

In case n=2, the results of Osserman [11] and Chern-Osserman [6] imply that $I_1 < \infty$ if and only if M^2 is conformally equivalent to a compact Riemann surface $\overline{M}^2$ punctured at a finite number of points and the Gauss map extends holomorphically over the punctures. This structure theorem serves as a model and motivates many of the problems in higher dimensions.

In the case of complex analytic varieties $M^n \to C^N$, these integrals have been studied by Griffiths [7], Langevin [10] and others. The integrals I_j occur as order functions in the value distribution theory of several complex variables [5], [9]; they and their averages over appropriate Grassmannians are major ingredients in the First Main Theorem. One may compute that

$$\mathrm{Tr}(\Lambda^{2k}R)dV = c_k(M) \wedge \phi^{n-k}$$

where $c_k(M)$ is the kth Chern class of M and ϕ is the standard Kahler form on C^N. Thus, the integrals I_k have a topological character in this case. Further, the integrand of I_k has the sign $(-1)^k$ pointwise. Griffiths (oral communication) has posed the following:

PROBLEM $P_k(n)$: Let $M^n \to C^N$ be a complete, non-singular, complex analytic n-dimensional variety in C^N. If

$$0 < I_k(M) = \lim_{r\to\infty} \frac{(-1)^k}{r^{n-2k}} \int_{M\cap B(r)} c_k(M) \wedge \phi^{n-k} < \infty \tag{2.4}$$

is M algebraic? (The converse is known to be true [7, §3c].)

Note that Stoll's theorem implies an affirmative answer for $P_0(n)$. Also, it is well known that $P_1(1)$ is true. In the next section, we show that the problem $P_k(n)$ for $n > k$, can be reduced to the problem $P_k(k)$; in other words, only the top dimensional problem remains to be solved.

There are a number of further curvature integrals of importance:

(1) $\int_{M^n} |K|$: the total absolute curvature in the sense of Chern-Lashof. Geometrically, this is the volume of the image of the Gauss map $G\colon S(N_M) \to S^{N-1}$.

(2) $\int_{M^n} |A|^n$: the L_1^n norm of the Gauss map. The exponent n is critical for the Sobolev embedding theorem. We refer to [3], [4] for results regarding this integral.

(3) Finally, there are integrals associated to other characteristic numbers of M (besides the Euler number) via the Chern-Weil theory; for instance Pontryagin numbers and signature. We do not study these integrals here.

Our basic problem is to understand what the finiteness of any of the above curvature integrals implies about the global geometry and topology of the minimal submanifold $M^n \to E^N$. In particular, if one of the curvature integrals $I_R(M) < \infty$, is M of class $\mathcal{T}(I)$? Does $I_R(M) < \infty$ imply M has a unique tangent cone at infinity? Is M of finite topological type?

Conversely, suppose that $M^n \in T(I)$ has a unique tangent cone $T = C(\Sigma)$ at infinity, where Σ is a smooth minimally immersed submanifold of $S^{N-1}(1)$, of multiplicity one away from the self-intersections. Then it is straightforward to see that all curvature integrals I_k, (1) and (3) above are finite.

Finally, we remark that one may use these integrals to study the behavior of tangent cones to a minimal variety with isolated singularities. Generally speaking, results one can prove for tangent cones at infinity of $M^n \to E^N$ will also hold in the local case of isolated singularities.

3. EXISTENCE OF TANGENT CONES. We begin this section with a discussion of complex varieties, in particular the problems P_k mentioned above. The following was done jointly with Nessim Sibony.

THEOREM 3.1. *Suppose that $P_k(k)$ is valid: that is if $M^n \to C^N$ is a complete, non-singular, complex analytic variety and $0 < I_k(M^k) < \infty$, then M^k is algebraic. Then $P_k(n)$ is valid for all $n > k$.*

PROOF. We need to prove, under the assumption that $P_k(k)$ is valid, that given a complex analytic variety $M^n \to C^N$, if $0 < I_k(M^n) < \infty$, then M^n is algebraic. Recall the Crofton formula for varieties in C^N (see e.g. p.479 of [7]);

$$\frac{1}{r^{n-2k}} \int_{M \cap B(r)} c_k(M) \wedge \phi^{n-k} = \int_{L \in G} \Big[\int_{M \cap B(r) \cap L} c_k(M \cap L)\Big] dL \tag{3.1}$$

where dL is the kinematic density of $G = G_{N,N-n+k}(C)$, the Grassmannian of affine (N-n+k) planes in C^N. By assumption, the left hand side is bounded, away from 0 and ∞, as $r \to \infty$; using Fubini's theorem, it follows that, for some $0 \in C^N$ and for almost all (N-n+k) planes H through 0,

$$\int_{M \cap H} c_k(M \cap H) < \infty .$$

Thus, since $P_k(k)$ is valid, $M \cap H$ is algebraic, for almost all H. It now follows from work of Gruman [8], that M itself is algebraic. □

COROLLARY 3.2. *Let $M^n \to C^N$ be a complete, non-singular complex analytic variety. Then $I_1(M^n) = \lim_{r \to \infty} \frac{(-1)}{r^{2n-2}} \int_{M \cap B(r)} c_1(M) < \infty$ if and only if M is algebraic.*

PROOF. It is well known that $P_1(1)$ is valid, i.e. if $M^1 \to C^N$ is a complex curve and $I_1(M^1) < \infty$, then M is algebraic. Thus, one direction of the corollary follows immediately from Theorem 3.1. For the converse, see § 2. □

For the case of minimal varieties M^n in E^N, it appears to be a more delicate question whether $I_1 < \infty$ implies that M^n $T(I)$ if $n > 2$. By scaling appropriately, it is sufficient to show that if $M^n \to B^N(1)$ is a minimal submanifold without boundary, then

$$\text{(3.2)} \qquad \mathrm{vol}(M) < C[1 + \int_M |A|^2].$$

This estimate is obvious if one knows pointwise that $|A|^2 > \varepsilon$ or $|A|^2 < \varepsilon$ (C then depends on ε). Note also that for arbitrary compact submanifolds of $B^N(1)$, (3.2) follows by integrating the identity $\Delta |X|^2 = \langle H, X \rangle + n$, where X is the position vector field. However, given C fixed, it is easy to construct embeddings of the ball $B^n \to B^N(1)$ which contradict (3.2) if $n > 2$. For instance, take a finite disjoint family of balls $B^n(1) \subset E^n$ and connect them by small tubes to obtain an n-ball in E^n; now 'fold' the balls over one another into $B^N(1)$ mapping the boundary to $S^{N-1}(1)$. One may arrange this so that the volume is arbitrarily large but I_1 remains bounded if $n > 2$. It is possible that there is a minimal embedding close to this embedding.

4. UNIQUENESS OF TANGENT CONES. Perhaps the simplest question one can pose regarding uniqueness of tangent cones and curvature integrals is the following. For simplicity, we deal with the codimension one case only.

CONJECTURE. Let $M^n \to E^{n+1}$ be a complete minimal immersion in E^{n+1} such that $0 < \int_M |K| < \infty$. Then M has a unique tangent cone at infinity.

We recall that $\int_M |K|$ is the n-dimensional volume of the image of the Gauss map $G(M) \to S^n$, counted with multiplicity. We do not even know if $M \in T(I)$ and the conjecture is still interesting if one assumes $M \in T(I)$. Notice that in case M^n is comples analytic, then $M \in T(I)$ implies M has a unique tangent cone at infinity. Similarly, one may conjecture that if $M^n \subset B^N(1)$ is a minimal variety with an isolated singularity at 0 such that $\int_M |K| < \infty$, then the tangent cone to M^n at 0 is unique. (In this case, the volume growth condition is automatic.)

We will prove the conjecture in two special cases. First, we need a definition. Let $G_{n-1}(TS^n)$ denote the oriented Grassman bundle of linear n-1 planes in TS^n. If $H_x \subset T_xS^n$ is an oriented hyperplane, let γ_H denote the unit speed geodesic on S^n such that $\gamma_H(0) = x$ and $(H_x, \frac{d}{dt}\gamma_H|_{t=0})$ gives the orientation of T_xS^n. Define

$$\tilde{G}: G_{n-1}(TS^n) \to G_{n-1}(TS^n)$$
$$\tilde{G}(H_x) = \left[\frac{d}{dt}\gamma_H(t)\Big|_{t=\pi/2}\right]^{\perp}.$$

If V is an oriented (n-1) varifold, we will call the oriented (n-1) varifold $(\tilde{G})_{\#}V$ the <u>polar variety</u> of V. The polar variety is <u>non-degenerate</u> if $(\tilde{G})_{\#}V \neq 0$. Note that if $V \to S^n$ is an immersed oriented (n-1)-submanifold the map $\tilde{G}$ corresponds to the usual Gauss map G, namely translation of the unit normal of V in S^n to the origin.

PROPOSITION 4.1. *Let $M^n \to E^{n+1}$ be a complete, oriented minimal immersion of class $\mathcal{T}(I)$ such that $\int_M |K| < \infty$. If the polar variety of a tangent cone at infinity to M is non-degenerate, then M has a unique tangent cone at infinity.*

PROOF. Let $T_1 = \lim_{j\to\infty} \frac{1}{r_j}[M \cap B(r_j)]$ and $T_2 = \lim_{j\to\infty} \frac{1}{s_j}[M \cap B(s_j)]$ be tangent cones at infinity of M with $T_i = C(\Sigma_i)$, where Σ_i is a stationary, oriented integral (n-1) varifold in S^n. Suppose that the polar variety to Σ_1 is non-degenerate. The set of tangent cones to M, considered as a subset of the space of (n-1) varifolds, is compact in the weak topology and has no isolated points, if it has at least two elements. By continuity, if $T_1 = T_2$, we may find a tangent cone, assumed to be T_2 for now, such that the polar variety of Σ_2 is also non-degenerate. Let $V_i = (\tilde{G})_{\#}\Sigma_i$; we claim that $V_1 = V_2$. To see this, note that $(\tilde{G})_{\#}V_i = (A)_{\#}\Sigma_i$, where A is the antipodal map of S^n. If $V_1 = V_2$, then $(A)_{\#}\Sigma_1 = (A)_{\#}\Sigma_2$ so that $\Sigma_1 = \Sigma_2$ which is impossible.

Consider the domains $G(M \cap A(r_j, s_j)) \subset S^n$, where $G: M^n \to S^n$ is the Gauss map and we assume that $r_j < s_j < r_{j+1}$. Note that as varifolds in S^n

$$\lim_{j\to\infty} G|_{M \cap S(r_j)} = \lim_{j\to\infty} \tilde{G}(M \cap S(r_j)) = \tilde{G}(\Sigma_1).$$

The same equation holds for $\{s_j\}$. Since $\tilde{G}(\Sigma_1) \neq \tilde{G}(\Sigma_2)$ as (n-1) varifolds, there is a $c > 0$ such that

$$\mathrm{vol}(G(M \cap A(r_j, s_j))) > c,$$

for all j. Summing on j implies that $\int_M |K| = \infty$. Thus we must have $T_1 = T_2$. $\square$

When the polar variety of a tangent cone is degenerate, on must argue differently.

THEOREM 4.2. *Let $M^n \to E^{n+1}$ be a complete, oriented, area-minimizing hypersurface such that $\int_M |K| < \infty$. Then M has a unique tangent cone at infinity.*

PROOF. The proof is patterned on Theorem 3.1 of [3]. It is well known that M^n is of class $\mathcal{T}(I)$, so that M has tangent cones at infinity. Suppose $T_1 = \lim_{i\to\infty} \frac{1}{r_i}[M \cap B(r_i)]$ in the weak topology on integral n-varifolds. Choose points $x_i \in M \cap S(\overline{R}_i)$, where $\overline{R}_i = \frac{1}{2}(r_i + r_{i+1})$ and consider the minimal submanifolds

$$V_i = \frac{1}{\tilde{R}_i}[M \cap A(r_i, r_{i+1}) - x_i] \cap B^N(1)$$

where $\tilde{R}_i = \frac{1}{2}(r_{i+1} - r_i)$. It follows from the well known monotonicity formula in E^N that $\mathrm{vol}(V_i)$ is uniformly bounded. Further, since V_i is area-minimizing V_i subconverges to an area-minimizing integral n-varifold V_∞. The regularity theory for such varifolds guarantees that V_∞ is a smooth n-manifold off a singular set $Z_2 \subset \mathrm{supp}\, V_\infty$ of codimension ≥ 7; the convergence $V_i \to V_\infty$ is C^∞ away from Z_2. Now, using the dilation invariance of the integral $\int_M |K|$, we have $\int_M |K| \to 0$ as $i \to \infty$, so that

$$\int_M |K| = 0.$$

It follows that the manifold $V_\infty - Z_2$ is ruled by straight line segments which are principal directions of the second fundamental form of $V_\infty - Z_2$. Thus $V_\infty - Z_2$ is completely determined by a full transversal to this ruling.

Let λ_0 denote the smallest eigenvalue, in absolute value, of the second fundamental form of $M^n \to E^{n+1}$ and let E_0 denote the corresponding eigenspace. Clearly, $\dim E_0$ is upper-semicontinuous, but is not necessarily constant on M. Nevertheless, there is a 1-dimensional distribution $E \subset E_0$ such that $E|_{V_i}$ converges smoothly on $V_\infty - Z_2$ to the distribution of the tangents of the ruling of $V_\infty - Z_2$. Further, when considered on $W_i = \frac{1}{r_{i+1}}[M \cap A(\frac{1}{2}r_{i+1}, 2r_{i+1})]$, the distribution E converges smoothly on $(T_1 - Z_1) \cap A(1/2, 2)$ to the tangents of the ruling of the cone T_1; here Z_1 denotes the singular set of T_1. Note that the Hausdorff dimensions of Z_1 and Z_2 are $\geq n-7$.

Consider the tangent cone $T_2 = C(\Sigma_2)$ associated to the sequence $\overline{R}_i$ (or a subsequence of $\{\overline{R}_i\}$). For a generic ray $L \subset \mathrm{supp} T_2$, the above argument implies that $L \subset \mathrm{supp} T_1$ and $T_{L \cap S^n}(\Sigma_1) = T_{L \cap S^n}(\Sigma_2)$. Thus, $T_1 = T_2$.

We may now repeat this argument inductively to prove that the tangent cone of M is unique. □

REMARK. By choosing sequences $\{r_i\} \to 0$ instead of $\{r_i\} \to \infty$, the same proof shows: if $M^n \subset B^{n+1}(1)$ is an area-minimizing variety in $B^{n+1}(1)$ with an isolated singularity at 0 such that $\int_M |K| < \infty$, then M has a unique tangent cone at 0.

We close with some open problems.

(1) Let $M^n \to E^{n+1}$ be a complete area-minimizing hypersurface. Is

$\int_M |K| < \infty$? Of course this is related to uniqueness of the tangent cones at infinity to M. Major progress on the uniqueness question in general has been made recently by Leon Simon.

(2) It is not difficult to show that if M^n is as above, then $I_1(M) < \infty$. Is $I_1 < \infty$ for a stable minimal hypersurface in E^{n+1}?

(3) It would be interseting to determine whether or not the condition $I_1 < \infty$ implies that $M \in T(I)$, for a complete minimal submanifold $M^n \to E^N$.

BIBLIOGRAPHY

!. W. K. Allard, "On the first variation of a varifold", Ann. of Math.,35, (1972), 417-491

2. M. T. Anderson, "Curvature estimates for minimal surfaces in 3-manifolds", Ann. Sci. Ecole Norm. Sup.,18:1 (1985)

3. M. T. Anderson, "The compactification of a minimal submanifold in Euclidean space by the Gauss map", (submitted to Acta Mathematica)

4. M. T. Anderson, "Local estimates for minimal submanifolds in dimensions greater than two", to appear in Proc. Symp. Pure Math., W. Allard and F. Almgren, Editors. Amer. Math. Soc.

5. J. Carleson and P. Griffiths, "The order functions for entire holomorphic mappings", in Value Distribution Theory, Dekker, New York, 1974, 225-248.

6. S.-S. Chern and R. Osserman, "Complete minimal surfaces in E^N", J. d'Analyse Math., 19, (1967), 15-34.

7. P. Griffiths, "Complex differential and integral geometry and curvature integrals associated to singularities of complex analytic varieties", Duke Math. J., 45:3 (1978), 427-512.

8. L. Gruman, "The area of analytic varieties in C^N", Math. Scand., 4 (1977), 365-397.

9. J. King and P. Griffiths, "Nevanlinna theory and holomorphic mappings between algebraic varieties", Acta Math., 130 (1973), 145-220.

10. R. Langevin, "Courbure et singularities complexes", Comm. Math. Helv., 54 (1979).

11. R. Osserman, "Global properties of minimal surfaces in E^3 and E^N", Ann. of Math., 80 (1964), 340-364.

12. W. Stoll, "The growth of the area of a transcendental analytic set", I and II, Math. Ann., 156 (1964), 47-78, 144-170.

13. H. Weyl, "On the volume of tubes", Amer. J. Math., (1939), 451-472.

DEPARTMENT OF MATHEMATICS
CALIFORNIA INSTITUTE OF TECHNOLOGY
PASADENA, CA 91125

Current Address:
Mathematical Sciences Research Institute
1000 Centennial Drive
Berkeley, CA 94720

Contemporary Mathematics
Volume 49, 1986

On the automorphism group of strictly convex domains in $\mathbb{C}^n$

J. Bland, T. Duchamp* and M. Kalka

Abstract

Let D and D' be two bounded, strictly convex domains in $\mathbb{C}^n$ with smooth boundaries. We use results of Lempert and Patrizio to show that if ϕ is a local biholomorphism between a neighborhood of $p \in D$ and a neighborhood of $p' = \phi(p) \in D'$ which preserves the Kobayashi metric "to infinite order" at p then ϕ extends to a biholomorphism between D and D'.

§1. Introduction

The purpose of this paper is to characterize the group Aut(D,p), of biholomorphisms of a bounded, strictly convex domain $D \subseteq \mathbb{C}^n$ which fix an interior point $p \in D$, in terms of infinitesimal data at p.

In [2] Lempert showed that if D has boundary class C^k, $k > 6$, then the Kobayashi metric on D, $\delta : D \times D \to \mathbb{R}$, is of class C^{k-4} away from the diagonal. More is true: Fix a point $p \in D$ and consider the function

$$(1.1) \qquad \delta_p : D \to \mathbb{R} : q \longmapsto \delta(p,q).$$

Unless D is biholomorphic to the ball, δ_p^2 will not be differentiable at p. However, denoting the space obtained by blowing up the point p by $\tilde{D}$ and the blow up of p by $\tilde{p} = \mathbb{P}(T(D)_p)$, we show that the function

$$(1.2) \qquad \sigma : \tilde{D} \to D \xrightarrow{\delta_p^2} \mathbb{R}$$

*On leave from the University of Washington.

is of class C^{k-4}. (Here $\breve{D} \to D$ denotes the blow down map.) The following definition therefore makes sense when $k = \infty$.

1.3. Definition. Let D and D' be two bounded, strictly convex domains in $\mathbb{C}^n$ with smooth boundaries. Let ϕ be a biholomorphism between a neighborhood of $p \in D$ and a neighborhood of $p' = \phi(p) \in D'$ and denote the induced biholomorphism between a neighborhood of $p \subseteq D$ and a neighborhood of $\breve{p}' \subseteq \breve{D}'$ by $\breve{\phi}$. Then ϕ is said to preserve the Kobayashi metric to infinite order at p if the two functions σ and $\sigma' \circ \breve{\phi}$ agree to infinite order all points of $\breve{p}$.

The main result of this paper is the following theorem.

1.4. Theorem. Let D and D' be bounded, strictly convex domains in $\mathbb{C}^n$ with smooth boundaries and let ϕ be a biholomorphism between neighborhoods of $p \in D$ and $p' = \phi(p) \in D'$ which preserves the Kobayashi metric to infinite order at p. Then ϕ extends to a biholomorphism between D and D'.

Let $D_r = \{q \in D \mid \delta_p(q) < r\}$ be the Kobayashi ball of radius r centered at p. The proof of Theorem 1.4 requires the following theorem, also proved in Section 3. It is similar to a recent result of Graham and Wu [1].

1.5. Theorem. Let D and D' be two bounded, strictly convex domains in $\mathbb{C}^n$ with boundaries of class C^k $(k \geq 6)$ and let D_r and D'_r be the Kobayashi balls of radius r about $p \in D$ and $p' \in D'$. Then every biholomorphism $\phi : D_r \to D'_r$ sending p to p' is the restriction of a biholomorphism between D and D'.

§2. The circular representation

In this section we wish to recall some results of Lempert [2]. Since Lempert proves more than he actually states, and since we need slightly sharper results than the actual statements of Theorems 2 and 3 of [2], we will give here a somewhat altered statement of Lempert's results. The treatment of the circular representation given by Patrizio in [3] is very close to ours - the main difference rests in our use of the blowup construction.

In addition to the notation of the introduction, denote the unit disc by $\Delta = \{\xi \mid |\xi| < 1\} \subseteq \mathbb{C}$ and the circle group by $T = \{e^{i\theta} \mid \theta \in \mathbb{R}\}$.

(2.1) Recall that a holomorphic map $f : \Delta \to D$ is said to be <u>extremal</u> with respect to $v \in T(D)_p$ (= tangent space to D at p), $v \neq 0$, if the following conditions are satisfied:

(i) $f(0) = p$,

(ii) $f'(0) = \lambda v$ for λ real and $\lambda > 0$

and

(iii) for every holomorphic $g : \Delta \to D$ with $g(0) = 0$, $g'(0) = \mu v$, $\mu > 0$ the inequality $\mu \leq \lambda$ holds.

Lempert shows [Theorem 2, Prop. 11'] that for each $v \neq 0$ <u>there is a unique holomorphic map</u> $f_v : \overline{\Delta} \to D$ <u>extremal with respect to</u> v <u>and that the map</u>

$$T(D)_p \setminus \{0\} \to C^{2+\frac{1}{2}}(\overline{\Delta}, D)$$

is of class C^{k-4}. (Here $C^{2+\frac{1}{2}}(\overline{\Delta}, D)$ denotes the space of maps h from $\overline{\Delta}$ onto D whose derivatives $h(\alpha)$ up to order 2 satisfy the inequality $|h^{(\alpha)}(z) - h^{(\alpha)}(w)| \leq C|z-w|^{1/2}$, $z,w \in \overline{\Delta}$.) Because the maps f_v are holomorphic on Δ the Cauchy integral formula can be used to show that <u>the map</u>

(2.2) $$\Psi : T(D)_p \setminus \{0\} \times \Delta \to D : (v, \xi) \longmapsto f_v(\xi)$$

<u>is of class</u> C^{k-4}. Lempert also shows in Theorem 2 that the map $\xi \longmapsto f(\frac{z}{|z|}\xi)$ is extremal for zv, $z \in \mathbb{C}\setminus\{0\}$. Hence, the equivariance identity

$$f_{zv}(\xi) = f_v(\frac{z}{|z|} v)$$

holds for all $v \in T(D)_p \setminus \{0\}$, $\xi \in \Delta$, $z \in \mathbb{P}\setminus\{0\}$.

There is a natural norm on the tangent space $T(D)_p$, which we will call the Kobayashi norm.

2.4. <u>Lemma</u>. The function $\| \ \| : T(D)_p \to \mathbb{R}$ defined by

$$f'_v(0) = v/\|v\| \ , \ v \neq 0$$

and

$$\|0\| = 0$$

is a norm on $T(D)_p$ and $\| \ \| : T(D)_p \setminus \{0\} \to \mathbb{R}$ is of class C^{k-4}.

Proof: First, note that the map $T(D)_p \setminus \{0\} \times \Delta \to \mathbb{C}^n : (v,\xi) \longmapsto f_v'(\xi)$ is of class C^{k-4} since

$$f_v'(\xi) = \frac{1}{2\pi i} \int_{|z|=1} \frac{\phi(v,z)}{(\xi-z)^2}\, dz$$

and ϕ of class C^{k-4} in v and continuous in z for $z \in \partial\Delta$. Hence,

$$\|v\| = v/f_v'(0)$$

is of class C^{k-4} for $v \neq 0$.

To prove that $\| \ \|$ is a norm we must show two things:
(i) $\|zv\| = |z|\|v\|$ for $z \in \mathbb{C}$ and (ii) $\|v+w\| \leq \|v\| + \|w\|$ for $v,w \in T(D)_p$.

Proof of (i): Given $z \neq 0$ it follows from formula (2.3) that $f_{zv}'(0) = \frac{z}{|z|} f_v'(0)$. Therefore, from the definition of $\| \ \|$, we have the equation

$$\frac{zv}{\|zv\|} = \frac{z}{|z|} \frac{v}{\|v\|}$$

from which (i) follows.

Proof of (ii): Assume that $p = 0 \in \mathbb{C}^n$, for convenience. By (i) we need only prove (ii) for v and w such that $v + w \neq 0$, $v \neq 0$, $w \neq 0$, $\|v\| + \|w\| = 1$. Set $g = \|v\|f_v + \|w\|f_w : \Delta \to D$. Observe that $g'(0) = v + w$ and $f_{v+w}'(0) = \frac{v+w}{\|v+w\|}$.

It then follows from the extremal property of f_{v+w} that the inequality $\frac{1}{\|v+w\|} \geq 1$ holds. But since $\|v\| + \|w\| = 1$, this can be written in the form $\|v+w\| \leq \|v\| + \|w\|$. □

Using the norm $\| \ \|$ it is possible to define a unit sphere and a unit ball in $T(D)_p \approx \mathbb{C}^n$ which we denote by S and B, respectively. Note, however, that in general B is <u>not</u> biholomorphic to the standard ball in $\mathbb{C}^n$; it is only a circular domain.

Consider now the restrictions of ϕ to $S \times \Delta$ and observe that because $\phi(v,0) = 0$ and $\xi \longmapsto \phi(v,\xi)$ is analytic for all $v \in S$ the map ϕ lifts to a C^{k-4} map into the blowup of D:

$$\breve{\phi} : S \times \Delta \to \breve{D} . \tag{2.5}$$

Define a circle action on $S \times \Delta$ by

$$e^{i\theta}\cdot(v,\xi) = (e^{i\theta}v,\ e^{-i\theta}\xi) \tag{2.6}$$

and let $\breve{B}$ denote the orbit space $S \times \Delta/T$. The reason for our notation is that $\breve{B}$ is naturally identified with the space obtained from B by blowing up the origin - in fact, the blow down map is induced by the map

$$S \times \Delta \to B$$
$$(v,\xi) \longmapsto \xi v .$$

By virtue of the identity (2.3) there is a commutative diagram

$$\begin{array}{ccc} \breve{B} & \xrightarrow{\ \breve{\Psi}\ } & \breve{D} \\ \big\downarrow & & \big\downarrow \\ B & \xrightarrow{\ \Psi\ } & D \end{array} \tag{2.7}$$

with $\breve{\Psi}$ of class C^{k-4} (vertical maps are blow downs). The map Ψ is called the circular representation (see [3]) of D at p and a holomorphic invariant of the pair (D,p).

2.8. Remarks (i) Lempert shows [[2],Theorem 3] that the map $\Psi : B\backslash\{0\} \to D\backslash\{p\}$ is a C^{k-4}-diffeomorphism. But since on the set

$$\breve{O} = \{[v,0] \in S \times \Delta/T\}$$

($[v,\xi]$ = orbit of $(v,0)$) the map $\breve{\Psi}$ is given by

$$\breve{\Psi}([v,0]) = [v] \in \mathbb{P}(T(D)_p) = \breve{p} \subseteq \breve{D}$$

and because $\xi \longmapsto \phi(v,\xi)$ is non-singular at $\xi = 0$ for all $v \neq 0$ it follows that $\breve{\Psi}$ is invertible along $\breve{O}$. Therefore, $\breve{\Psi}$ <u>is a C^{k-4}-diffeomorphism.</u>

(ii) $\breve{B}$ can - and should - be thought of as the unit disc bundle of the canonical line bundle $L \xrightarrow{\pi} \mathbb{P}(T(D)_p)$ equipped with the Hermitian metric induced by $\| \ \|$. The map $\breve{\Psi}$ then maps each fiber of $\breve{B} \longrightarrow \mathbb{P}(T(D)_p)$ biholomorphically onto the image of an extremal disc and furnishes a non-singular foliation $F_{\breve{D}}$ of $\breve{D}$ by holomorphically embedded discs. We call $F_{\breve{D}}$ the <u>Lempert foliation</u>. Note that $F_{\breve{D}}$ is transverse to the hypersurface $\breve{p} \cong \mathbb{P}(T(D)_p)$.

(iii) The function $\breve{B} \to B \xrightarrow{\| \ \|^2} \mathbb{R}$ is of class C^{k-4} because it is induced by the function $S \times \Delta \to \mathbb{R} : (v,\xi) \longmapsto |\xi|^2$. In fact, it is real analytic on each fiber of π. Consequently the function

$$\tau : D \xrightarrow{\breve{\Psi}^{-1}} \breve{B} \longrightarrow B \xrightarrow{\| \ \|^2} \mathbb{R}$$

is also of class C^{k-4} and analytic on each extremal disc centered at p. Since the function δ_p of (1.1) is shown in [2] to be given by the formula

$$\begin{aligned}\delta_p(q) &= \frac{1}{2} \log \frac{1 + \sqrt{\tau(q)}}{1 - \sqrt{\tau(q)}} \\ &= \sqrt{\tau(g)}(1 + \tau(q)/3 + \tau(q)^2/5 + \tau(q)^3/7 + \ldots)\end{aligned}$$

it follows that <u>the function σ of (1.2) is of class C^{k-4} and real analytic on each disc of $F_{\breve{D}}$.</u>

(iv) Given $0 < r < \infty$ and $0 < \varepsilon < 1$, related by $r = \frac{1}{2}\log(\frac{1+\varepsilon}{1-\varepsilon})$, the domain $D_r = \{q \in D \mid \delta_p(q) < r\} = \{q \in D \mid \tau(q) < \varepsilon\}$ is convex with boundary of class C^{k-4}. Lempert shows [2, top of page 464] that there is a circular representation $\Phi_r : B \to D_r$ for D_r. The norms $\| \ \|$ and $\| \ \|_r$ on

$T(D)$ and $T(D)$ are related by the equation

$\|v\|_r = \|v\|/\varepsilon$ so $B_r = \{v \in T(D)_p \mid \|v\| < \varepsilon\}$ and the diagram

$$\begin{array}{ccc} B_r & \xrightarrow{\Psi_r} & D_r \\ \cap & & \cap \\ B & \xrightarrow{\Psi} & D \end{array}$$

commutes.

§3. Proofs of Theorems 1.4, 1.5

The proofs of Theorems 1.4 and 1.5 depend on choosing a good coordinate system in which to represent certain tensors. Such coordinates are furnished by the circular representation $\check{\Psi}$. Choose a point $q \in \mathbb{P}(T(D)_p)$ and a sufficiently small neighborhood V of q so that the restriction $\check{B}_{|V} \to V$ is trivial and V is biholomorphic to a domain $U \subseteq \mathbb{C}^{n-1}$. Then there is a C^{k-4} diffeomorphism

$$\text{(3.1)} \qquad \begin{array}{c} \phi : U \times \Delta \approx B/V \xrightarrow{\check{\Psi}} \check{D} \\ (w,\xi) \longmapsto \phi(w,\xi), \end{array}$$

onto an open set in $\check{D}$ such that the maps $\Delta \to D : \xi \longmapsto \phi(w,\xi)$ are extremal discs. Let $w = (w^1,\dots,w^{n-1})$, $w^\alpha = u^{2\alpha-1} + iu^{2\alpha}$, $\alpha = 1,\dots,n-1$ and $w^n = u^{2n-1} + iu^{2n} = \xi$ be the coordinates on $\check{D}$ induced by the map ϕ. Such coordinates are said to be adapted to the foliation $F_{\check{D}}$.

Remark: In general these are not holomorphic coordinates.

3.2. Lemma. Let T be a real analytic tensor defined on an open set of $\check{D}$. Then in adapted coordinates

$$T = T^{\alpha_1\dots\alpha_s}_{\beta_1\dots\beta_t} \frac{\partial}{\partial u^{\alpha_1}} \otimes \dots \otimes \frac{\partial}{\partial u^{\alpha_s}} \otimes du^{\beta_1} \otimes \dots \otimes du^{\beta_t}$$

the functions $T^A_B = T^{\alpha_1\dots\alpha_s}_{\beta_1\dots\beta_t}$ are real analytic functions of u^{2n-1} and u^{2n}.

Proof. Let $y = (y^1, \ldots, y^{2n})$ be real analytic coordinates on $\tilde{D}$. In these coordinates, the components of T are real analytic in y. Since the change of coordinate function $y = \phi(w)$ is real analytic in u^{2n-1} and u^{2n} so is its derivative $d\phi$*. Therefore, the change of coordinates formula for the components of T involves only functions which are analytic in u^{2n-1} and u^{2n} and therefore T^A_B are analytic in u^{2n-1} and u^{2n}. □

3.3 Proof of Theorem 1.5. Let Ψ_r and Ψ'_r be the circular representations of D_r and D'_r centered about p and p'. Because the circular representation is a holomorphic invariant the following diagram commutes

(3.4)
$$\begin{array}{ccccc} T(D_r)_p & \supseteq B_r & \xrightarrow{\Psi_r} & D_r \\ & \Big\downarrow d\phi_p & & \Big\downarrow \phi \\ T(D'_r)_{p'} & \supseteq B'_r & \xrightarrow{\Psi'_r} & D'_r \end{array}$$

By Remarks 2.8(iv) the linear map $d\phi_p : T(D)_p \to T(D')_{p'}$ maps $B = \{v \mid \|v\| < 1\}$ to $B' = \{v' \mid \|v'\|' < 1\}$ and diagram (3.4) extends to a commutative diagram

$$\begin{array}{ccc} B & \xrightarrow{\Psi} & D \\ \Big\downarrow d\phi_p & & \Big\downarrow \Phi \\ B' & \xrightarrow{\Psi'} & D' \end{array}$$

thus defining a C^{k-4}-diffeomorphism Φ which is a biholomorphism on D_r. Note also that if $f : \Delta \to D$ is an extremal map with respect to $v \in T(D)_p$, then $\Phi \circ f : \Delta \to D'$ is extremal with respect to $d\Phi_p(v)$.

---

* We are using here the standard fact that if $g(x,y)$ is a continuously differentiable function of x and y and real analytic in y, then the partial derivative, $\frac{\partial g}{\partial x}$, is also real analytic in y.

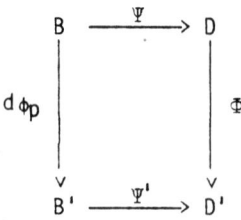

Now let $\chi : U \times \Delta \to \breve{D}$ be a coordinate plot adapted to $F_{\breve{D}}$. Then by the above remarks $\chi' = \breve{\Phi} \circ \chi : U \times \Delta \to \breve{D}'$ is a coordinate chart adapted to $F_{\breve{D}'}$. Here $\breve{\Phi} : \breve{D} \to \breve{D}'$ is the extension of Φ to the blowups $\breve{D}$ and $\breve{D}'$ - it exists since Φ is a biholomorphism near p.

Since sets of the form $\chi(U \times \Delta)$ cover $\breve{D}$, to show that $\breve{\Phi}$ (and hence Φ) is holomorphic it is sufficient to prove the equality $\chi^*(J) = \chi'^*(J')$ between the pull backs of the complex structure tensors $J \in \Gamma(T^*(D) \otimes T(D))$ and $J' \in \Gamma(T^*(D') \otimes T(D'))$ on $U \times \Delta$.

Both J and J' are analytic tensors. Therefore, by Lemma 3.2 the components of $\chi^*(J)$ and $\chi'^*(J')$ are analytic in the coordinate u^{2n-1} and u^{2n}. But $\chi^*(J) = \chi'^*(J')$ on the set $U \times \Delta_r$ where $\Delta_r = \{\xi = u^{2n-1} + u^{2n} \in \Delta \mid \frac{1}{2} \log(\frac{1 + |\xi|}{1 - |\xi|}) < r\}$. Therefore $\chi^*(J) = \chi'^*(J')$ on all of $U \times \Delta$. □

3.5. Proof of Theorem 1.4. For sufficiently small $r > 0$, ϕ is a biholomorphism between the Kobayashi ball of radius r centered at p, D_r, and an open set $U \subseteq D'$. The induced biholomorphism on blowups will then be written $\breve{\phi} : \breve{D}_r \to \breve{U}$. The Lempert foliation of $\breve{D}$ then restricts to a foliation $F_{\breve{D}_r}$ of $\breve{D}_r$ by discs and the Lempert foliation $F_{\breve{G}'}$ restricts to a foliation $F_{\breve{U}}$ of $\breve{U}$. We claim that $\breve{\phi}$ maps each leaf of $F_{\breve{D}_r}$ into a leaf of $F_{\breve{U}}$.

Given the claim, the proof of 1.4 follows easily. For since the functions τ and τ' are real analytic on the leaves of $F_{\breve{D}}$ and $F_{\breve{D}'}$, respectively, it follows that τ and $\tau' \circ \breve{\phi}$ are real analytic on the leaves of F_{D_r} and therefore since they have the same Taylor series at one point of each leaf (the hypersurface $\breve{p}$ is transverse to the leaves of F_{D_r}) they agree on all of $\breve{D}_r$. It follows that the set U is just the Kobayashi ball, D'_r, in D' of radius r centered at p'. Now use Theorem 1.5 to conclude that ϕ extends to a biholomorphism between D and D'.

It remains to prove the claim. We begin by recalling some properties of the foliation $F_{\breve{D}}$ and the function τ (the discussion applies equally well to $F_{\breve{D}'}$ and τ). Choose an arbitrary point $q \in \breve{p} \subseteq \breve{D}$ and let $(z,w) \in \mathbb{C} \times \mathbb{C}^{n-1}$

be local holomorphic coordinates of $\check{D}$ centered at q with the hypersurface $\check{p}$ given by the equation $z = 0$. Then since the leaves of $F_{\check{D}}$ are transverse to $\check{p}$ the leaf of $F_{\check{D}}$ through q, denoted by L_q, is of the form

$$L_q = \{(z,f(z) \mid zV\}$$

where $f : \mathbb{C} \supseteq V \to \mathbb{C}^{n-1}$ is a holomorphic map defined on a neighborhood of 0 and $f(0) = 0$.

By [2, Theorem 4] (or from the results of section 2 above) the function τ is locally of the form

$$\tau(z,w) = |z|^2 h(z,w) \tag{3.6}$$

where h is a smooth function such that $h(0,w) > 0$*. It follows that the (1,1)-form

$$\omega_D = i\partial\bar{\partial} \log \tau \tag{3.7}$$

is smooth and that if we write ω_D in the form

$$\omega_D = A_0(z,w)dz \wedge d\bar{z} + \sum_{\alpha=1}^{n-1} A_{\bar{\alpha}}(z,w)dz \wedge d\bar{w}^\alpha + A_\alpha(z,w)d\bar{z} \wedge dw^\alpha + \sum_{\alpha,\beta} B_{\alpha\bar{\beta}}(z,w)dw^\alpha \wedge d\bar{w}^\beta \tag{3.8}$$

then the Taylor series at $(0,0)$ of the component functions, A_i, B_{ij} are completely determined by the Taylor series τ at $(0,0)$.

Lempert also shows that ω_D has rank $n-1$ and that if $X \in T^{1,0}(\check{D})$ is tangent to a leaf of $F_{\check{D}}$ then the equation

$$Z \quad \omega_D = 0 \tag{3.9}$$

holds. The tangent vector field to L_q,

$$Z = \frac{\partial}{\partial z} + \sum_{\alpha=1}^{n-1} \frac{\partial f^\alpha}{\partial z}(z) \frac{\partial}{\partial w^\alpha}. \tag{3.10}$$

---

*The function τ is the composition of the blowdown map $\check{D} \to D$ with the square of the function $u(x)$ defined in [2].

therefore, satisfies the rank $(n-1)$ system of equations

$$(3.11)\qquad \begin{aligned} &A_0(z,f(z)) + \sum_{\alpha=1}^{n-1} A_{\bar{\alpha}}(z,f(z))\frac{\partial f^\alpha}{\partial z}(z) = 0 \\ &A_{\bar{\beta}}(z,f(z)) + \sum_{\alpha=1}^{n-1} B_{\alpha\bar{\beta}}(z,f(z))\frac{\partial f^\alpha}{\partial z}(z) = 0 \qquad \beta = 1,2,\ldots,n-1. \end{aligned}$$

In particular, by a linear change of coordinates we may arrange for ω_D to satisfy the condition

$$(3.12)\qquad \omega_D(0,0) = i \sum_{\alpha=1}^{n-1} dw^\alpha \wedge d\bar{w}^\alpha$$

and thereby insure that the $(n-1) \times (n-1)$ matrix $[B_{\alpha\bar{\beta}}]$ is invertible near $(0,0)$. Letting $[B^{\alpha\bar{\beta}}]$ denote its inverse, the second set of equations (3.11) can be written in the form

$$(3.13)\qquad \frac{\partial f^\alpha}{\partial z}(z) = -\sum_{\beta=1}^{n-1} B^{\alpha\bar{\beta}}(z,f(z))\, A_{\bar{\beta}}(z,f(z)) \qquad \alpha = 1,2,\ldots,n-1.$$

By repeated differentiation of (3.13) at $z = 0$ we find that the holomorphic map f is completely determined by the Taylor series of the function τ at the point $(0,0)$.

Consider now the inverse image under $\check{\phi}$ of the leaf $L'_{q'}$ of $F_{\check{D}'}$ through $q' = \check{\phi}(q)$. Since $F_{\check{D}'}$ is transverse to $\check{p}'$ and since $\check{\phi}$ induces a biholomorphism between $\check{p}$ and $\check{p}'$ we have the local representation

$$\check{\phi}^{-1}(L'_{q'}) = \{(z,f'(z)) \mid z \in V\}$$

for $f' : \mathbb{C} \supset V' \to \mathbb{C}^{n-1}$ holomorphic. The above analysis (now applied to the function $\tau'\circ\check{\phi}$ and the form $\check{\phi}^*(i\partial\bar{\partial}\log\tau')$ shows that f' is completely determined by the Taylor series of $\tau'\circ\check{\phi}$ at $(0,0)$ via equations similar to (3.13). But the Taylor series at $(0,0)$ of τ and $\tau'\circ\check{\phi}$ coincide, and therefore so do the Taylor series at $(0,0)$ of the coefficient functions of $i\partial\bar{\partial}\log\tau$ and $i\partial\bar{\partial}\log(\tau'\circ\check{\phi}) \equiv \check{\phi}^*(i\partial\bar{\partial}\log\tau')$. It follows then that $f' \equiv f$ and consequently that $\check{\phi}(L_q)$ is contained in $L'_{q'}$. □

Bibliography

[1] I. Graham and H. Wu, Characterizations of the unit ball B^n in complex Euclidean space, (preprint).

[2] L. Lempert, La métrique de Kobayashi et la representation des domains sur la boule, Bull. Soc. Math. France 109 (1981), 427-474.

[3] G. Patrizio, Parabolic exhaustions for Strictly Convex Domains, Manuscripta Math. 47 (1984), 271-309.

Department of Mathematics
Tulane University

Contemporary Mathematics
Volume **49**, 1986

Inequality Between Chern Numbers of Singular Kähler Surfaces and Characterization of Orbit Space of Discrete Group of SU(2,1)

by S.Y. Cheng and S.T. Yau

In 1976, the second author [Ya1] proved the Calabi conjecture and demonstrated an inequality between Chern numbers $(-1)^n C_2 C_1^{n-2}$ and $\frac{n(-1)^n C_1^n}{2(n+1)}$ for algebraic manifolds with ample canonical line bundle. He also proved that algebraic manifold which is covered by the ball is characterized among manifolds with ample canonical line bundle by the equality of the above Chern numbers. At around the same time, Miyaoka [Mi1], extending the method of Bogomolov, was able to obtain the inequality for $n = 2$. In this case, Miyaoka needed only to assume the canonical line bundle to be positive outside some divisors.

In 1977, the authors succeeded in generalizing the Calabi conjecture to non-compact manifolds. Part of this generalization was published in [C-Y]. Part of it was announced in [Ya2]. For quasiprojective manifolds, one can obtain similar inequalities between Chern numbers except that one has to replace the Chern classes by logarithmic Chern classes. These inequalities were also obtained by Sakai [Sa] and Miyaoka around the same time. One of the major arguments we

0271-4132/86 $1.00 + $.25 per page

used in [C-Y] was the observation that most arguments that one needs to prove the existence of Kähler-Einstein metrics with negative scalar curvature are basically local. In fact, most of the arguments work if one can resolve singularities by holomorphic maps from nice manifolds onto the singularities so that the pulled back metric is good. In particular, we know how to study V-manifolds and find the inequality between Chern numbers of these manifolds.

In 1980, the second author gave a course on the Kähler-Einstein metric in Princeton and presented some of the results in the course. We wish to thank those participants in the course, including Burns and Noguchi, for their enthusiasm and their help. R. Klotz, who was a student of the second author, also wrote a thesis on the Kähler-Einstein metrics on surfaces with singularity. It should be noted that the study of surfaces with singularity enables one to prove the inequality $3C_2 \geqslant C_1^2$ for surfaces of general type and that $3C_2 = C_1^2$ implies the surface is covered by the ball. This was also observed by Miyaoka [Mi2] independently.

In this paper, we only present our results in two dimension. The arguments can be generalized to higher dimension and will be published elsewhere. In the case of two dimension, a discrete group acting on the ball has two types of singularities in the orbit space. One type is isolated singularities and the other type is fixed curves of some element of the group. Theorem 3.1 can be considered as the main result of this paper.

§1. General Existence Theorem

Let M be an n-dim complex manifold. Let A_i, $1 \leqslant i \leqslant a$, C_j, $1 \leqslant j \leqslant c$, D_k, $1 \leqslant k \leqslant d$, be divisors on M such that $S = \bigcup_{i=1}^{a} A_i \cup \bigcup_{j=1}^{c} C_j \cup \bigcup_{k=1}^{d} D_k$ is a divisor with normal crossings. Let $A = \bigcup_{i=1}^{a} A_i$, $C = \bigcup_{j=1}^{c} C_j$, $D = \bigcup_{k=1}^{d} D_k$. The notation is set up so that C represents the infinities, D represents local branching and A represents global branching. We impose the following conditions on A, C, D.

1. C is disjoint from $A \cup D$.

2. Each A_i is a connected component and there exists an open neighborhood U_i containing A_i and disjoint from C, a complex manifold $\tilde{U}_i$, a surjective proper holomorphic map $\pi_i : \tilde{U}_i \to U_i$, subvarieties $\tilde{A}_i = \pi_i^{-1}(A_i)$, $\tilde{D}_i = \pi_i^{-1}(D_i)$, a finite group of biholomorphic transformation G_i on $\tilde{U}_i$ such that its action commutes with π_i and G_i acts on $\tilde{U}_i \backslash (\tilde{A}_i \cup \tilde{D}_i)$ as a covering transformation group of $\pi_i : \tilde{U}_i \backslash (\tilde{A}_i \cup \tilde{D}_i) \to U_i \backslash (A_i \cup D_i)$.

3. For each $x \in D \backslash \bigcup_{i=1}^{a} U_i$ there is an open set V_x which contains x and disjoint from $A \cup C$, a complex manifold $\tilde{V}_x$ and a surjective proper holomorphic map $p_x : \tilde{V}_x \to V_x$, a subvariety $\tilde{D}_x = p_x^{-1}(D)$, a finite group of biholomorphic transformation H_x on $\tilde{V}_x$ such that its action commutes with p_x and H_x acts on $\tilde{V}_x \backslash \tilde{D}_x$ as a covering transformation group of $p_x : \tilde{V}_x \backslash \tilde{D}_x \to V_x \backslash D$.

4. If $V_x \cap V_{x'} \neq \emptyset$, then there is a biholomorphic map $\psi_{x'x} : p_i^{-1}(V_x \cap V_{x'}) \to p_{x'}^{-1}(V_x \cap V_{x'})$ which commutes with the group action H_x and $H_{x'}$. If $V_x \cap U_i \neq \emptyset$, then there is a biholomorphic map $\psi_{ix} : p_x^{-1}(V_x \cap U_i) \to \pi_i^{-1}(V_x \cap U_i)$ which commutes with the group action H_x and G_i .

We shall use the abbreviation $\{\pi_i, p_x\}$ to denote the above mentioned structure on M .

Remark 1.1 : A typical case of local branching is as follows: Let each D_k be irreducible. Let m_k be a positive integer associated with D_k . Let $x \in D$ and suppose that $x \in D_{k_1} \cap \cdots \cap D_{k_\ell}$ and $x \notin D_k$ if $k \notin \{k_1, ..., k_\ell\}$. Then locally there exists a coordinate system $Z_1, ..., Z_n$ so that $D_{k_i} = \{Z_i = 0\}$.

Than a local branching is given by $p_x(Z_1, ..., Z_n) = (Z_1^{m_1}, ..., Z_\ell^{m_\ell}, Z_{\ell+1}, ..., Z_n)$.

Definition 1.1 : A hermitian metric on $M\backslash S$ is said to be complete and compatible with $\{\pi_i, p_x\}$ iff: (i) There exists an open set W so that W contains C and disjoint from $A \cup D$ so that dS^2 is a complete metric on $W\backslash C$ with boundary ∂W and is "good" in the sense of Mumford [Mu] near C ; (ii) $\pi_i^*(dS^2)$ can be extended to $\tilde{U}_i$ smoothly as a complete metric with boundary $\partial\tilde{U}_i$; (iii) For $x \in D\backslash\bigcup_{i=1}^{a} U_i$, $p_x^*(dS^2)$ can be extended smoothly to $\tilde{V}_x$ as a complete metric on $\tilde{V}_x$ with boundary ∂V_x .

The purpose of this section is to formulate and construct a complete Kähler-Einstein metric compatible with $\{\pi_i, p_x\}$ under certain assumptions. The idea is to deform a background metric by solving a complex Monge-Ampère equation. The metric will be obtained as the negative of the Ricci tensor of a volume form on $M\backslash S$. The volume form is constructed by patching local volume forms satisfying the following conditions:

Condition 1 : There exists an open set W containing C and disjoint from $A \cup D$ and a volume form dV_C on $W\backslash C$ so that its Ricci tensor is negative definite and defines a complete Kähler metric on $W\backslash C$ with boundary ∂W and is a "good" metric in the sense of Mumford near C .

Condition 2 : There exists a volume form dV_D on $M\backslash D$ with negative definite Ricci tensor. We assume that $p_x^*(dV_D)$ extends smoothly to $\tilde{V}_x$ with negative definite Ricci tensor and defines a complete Kähler metric on $\tilde{V}_x$ with boundary $\partial\tilde{V}_x$. For $i \in \{1, ..., a\}$ we assume that $\pi_i^*(dV_D)$ extends smoothly to $\tilde{U}_i\backslash\tilde{A}_i$ as a volume form with negative definite Ricci tensor and that on a compact neighborhood of $\tilde{A}_i$ the Ricci tensor is bounded away from zero uniformly. Moreover, we assume that $\pi_i^*(dV_D)$ can be extended across $\tilde{A}_i$ with negative definite Ricci tensor in the generalized sense.

Condition 3 : There exists a Kähler metric dS_i^2 which is complete on $\tilde{U}_i$ with boundary $\partial\tilde{U}_i$ and is invariant under the group G_i .

Theorem 1.1 : Suppose Condition 1 - Condition 3 are satisfied on M . Then there exists a smooth volume form dV on $M\backslash S$ such that it has negative definite Ricci tensor and the Kähler metric thus obtained is a complete Kähler metric compatible with $\{\pi_i, p_x\}$.

Proof : There exists a partition of unity of nonnegative functions on M ,

ϕ, ϕ_C, ϕ_i, $1 \leqslant i \leqslant a$, such that $\phi + \phi_c + \sum_{i=1}^{a} \phi_i \equiv 1$, supp $(\phi_C) \subseteq W$ and $\phi_i \equiv 1$ in a neighborhood of C, supp $(\phi_i) \subseteq U_i$, $\phi_i \equiv 1$ in a neighborhood of A_i. Then $dV_D = \phi dV_D + \phi_C dV_D + \sum_{i=1}^{a} \phi_i dV_D$. By assumption, $\pi_i^*(dV_D)$ extends smoothly to $\tilde{U}_i \backslash \tilde{A}_i$ and can be extended in the generalized sense to have negative Ricci tensor. Using the invariant metric dS_i^2 we can mollify $\pi_i^*(dV_D)$ with respect to this metric and obtain a volume form with negative Ricci tensor and approximates $\pi_i^*(dV_D)$ up to high order on a compact set containing $\pi_i^*(\text{supp } \phi_i)$. We then push this volume form down to supp ϕ_i via π_i and denote it by dV_D^i. Then $\phi_i dV_D^i$ is a smooth form on $U_i \backslash (A_i \cup D)$. Clearly, the volume form $\phi\, dV_D + \phi_C\, dV_D + \sum_{i=1}^{a} \phi_i\, dV_D^i$ defines a volume form on $M \backslash (A \cup D)$ with negative Ricci tensor and its pullback under p_x or π_i can be extended smoothly. Now by choosing a large enough constant α, $dV = \alpha(\phi dV_D + \phi_C dV_D + \sum_{i=1}^{a} \phi_i\, dV_D^i) + \phi_C\, dV_C$ satisfies the required properties.

Proposition 1.2 : Suppose that $C = \emptyset$ and each D_i is irreducible. Let m_i be a position integer associated to D_i. Suppose that $K + \sum_{i=1}^{a} \frac{m_i - 1}{m_i} D_i > 0$ on M, where K is the canonical class of M. Suppose that $\tilde{U}_i$ is a pseudoconvex manifold with ample canonical class and that $\tilde{A}_i$ can be blown down to a subvariety of codimension greater than or equal to two. We assume further that if $D_j \cap U_i \neq \emptyset$ then the subgroup of G_i which fixes some point of $\pi_i^{-1}(D_j)$ is of order m_j. Then there is a volume form on $M \backslash (A \cup D)$ with negative Ricci tensor and the Kähler metric it defines is a complete Kähler metric compatible with $\{\pi_i, p_x\}$, where p_x is given by the Remark 1.1.

Proof : Let L_i be the line bundle determined by D_i and let S_i be a holomorphic section of L_i which vanishes exactly on D_i. Let $h_i(\cdot\,,\cdot)$ be a hermitian metric on L_i. Let $d\Omega$ be a volume form on M. Then let

$$dV_D = \prod_{i=1}^{d} \left(h_i(S_i, S_i)^{-2}\right)^{\frac{m_i-1}{m_i}} \times \left(1 - (h_i(S_i, S_i))^{\frac{2}{m_i}}\right)^{-2} \times d\Omega .$$

A simple computation shows that we may choose h_i so that dV_D has negative definite Ricci tensor on $M \backslash D$. Clearly, $p_x^*(dV_D)$ can be extended smoothly across $\tilde{D}_x$. Now for $\tilde{U}_i$ there is a complete Einstein-Kähler metric by assumption and is hence invariant under G_i. Since $\tilde{A}_i$ can be blown down to a subvariety of codimension greater than or equal to two one can then extend $\pi_i^*(dV_D)$ across $\tilde{A}_i$ in the generalized sense. The proposition then is a consequence of Theorem 1.1.

Now suppose that dV is a volume form on $M \backslash S$ with negative definite Ricci tensor such that the Kähler metric it defines is a complete Kähler metric compatible with $\{\pi_i, p_x\}$. Let dS^2 denote the Kähler metric. Under local coordinates $Z_1, \ldots, Z_n$ it can be expressed as $\Sigma\, g_{k\bar{}}\, dZ_k\, d\bar{Z}$. Then its volume form can be written as $e^f dV$. Note that f is a smooth and bounded function on $M \backslash S$ such that $\pi_i^*(f)$ can be extended as a smooth function on $\tilde{U}_i$ and $p_x^*(f)$ can be extended as a smooth function on $\tilde{V}_x$. Moreover, the derivatives of f with respect to dS^2 is bounded in a neighborhood of C. So we propose to solve:

$$(1.1)\quad \det(g_{k\bar{\ell}} + u_{k\bar{\ell}}) = e^{u+F} \det(g_{k\bar{\ell}}),$$

such that $(g_{k\bar{\ell}} + u_{k\bar{\ell}})$ is positive definite and uniformly equivalent to $(g_{k\bar{\ell}})$ on $M\backslash S$. So if we choose F to be $-f$ then $\Sigma(g_{k\bar{\ell}} + u_{k\bar{\ell}})\, dZ_k\, d\bar{Z}_\ell$ defines a complete Kähler-Einstein metric on $M\backslash S$ compatible with $\{\pi_i, p_x\}$. The method is basically the same as in [C-Y] but we have to adapt the tools for the present situation.

First of all we have to introduce some Banach spaces of functions on $M\backslash S$.

Let k be a nonnegative integer and $\delta \in (0,1)$. We say that u is a $C^{k,\delta}$ function on $M\backslash S$ compatible with $\{\pi_i, p_x\}$ whenever u is $C^{k,\delta}$ on $M\backslash S$ and that $\pi_i^*(u)$, $p_x^*(u)$ can be extended as a $C^{k,\delta}$ function to $\tilde{U}_i$ and $\tilde{V}_x$ respectively. Then we will introduce a $C^{k,\delta}$ norm for $C^{k,\delta}$ functions. Note that dS^2 has bounded curvature by assumption and hence there exists $\epsilon_0 > 0$ such that there are no conjugate points on a geodesic segment with length less than ϵ_0. In fact, we can assume ϵ_0 is so small that $\bigcup_{i=1}^{a} \pi_i(\tilde{U}_i(\epsilon_0)) \cup \bigcup_x p_x(\tilde{U}_x(\epsilon_0))$ is an open set containing $A \cup D$, whose $\tilde{U}_i(\epsilon_0)$, $\tilde{V}_x(\epsilon_0)$ consists of points whose distances to $\partial\tilde{U}_i$ and $\partial\tilde{V}_x$ is more than ϵ_0 with respect to the metric $\pi_i^*(dS^2)$ and $p_x^*(dS^2)$ respectively. For $x \in M\backslash S$ such that x is ϵ_0 away from $A \cup D$ let $\|u\|_{k,\delta}^{B_{\epsilon_0}(x)}$ be the $C^{k,\delta}$-norm on the ball of radius ϵ_0 on $T_x(M)$ of the pullback of u with respect to the exponential map centered at x. For $y \in \tilde{V}_x$ or $y \in \tilde{U}_i$ such that its distance to the boundary is at least ϵ_0 we define $\|u\|_{k,\delta}^{\tilde{B}_{\epsilon_0}(y)}$ to be the $C^{k,\delta}$-norm on the ball of radius ϵ_0 on $T_y(\tilde{V}_x)$ or

$T_y(\tilde{U}_i)$ of the pullback of $p_x^*(u)$ or $\pi_i^*(u)$ with respect to the exponential map centered at y. With the above mentioned notation we define:

$$\|u\|_{k,\delta;\epsilon_0} = \max\Big(\sup\{\|u\|_{k,\delta}^{B_{\epsilon_0}(x)} : d(x, A \cup D) > \epsilon_0\},$$

$$\sup\{\|u\|_{k,\delta}^{\tilde{B}_{\epsilon_0}(y)} : y \in \tilde{V}_x \text{ or } y \in \tilde{U}_i \text{ such that}$$

$$d(y, \partial\tilde{V}_x) > d(y, \partial\tilde{U}_i) > \epsilon_0\}$$

We want to point out that $\|u\|_{k,\delta;\rho}$ is equivalent to $\|u\|_{k,\delta;\epsilon_0}$ for any $\rho \in (0, \epsilon_0)$. This fact allows us to use interior Schauder estimates. Then $C^{k,\delta}(M,\{\pi_i, p_x\})$ denotes the Banach space of $C^{k,\delta}$ functions on $M\backslash S$ which is compatible with $\{\pi_i, p_x\}$ and $\|u\|_{k,\delta;\epsilon_0}$ is finite. Now we can formulate the equation we want to solve more rigorously. For integer $k \geqslant 5$, $\delta \in (0, 1)$ and $F \in C^{k-2,\delta}(M,\{\pi_i, p_x\})$ we want to find $u \in C^{k,\delta}(M,\{\pi_i, p_x\})$ such that

$$(1.2)\quad \begin{cases} \det(g_{k\bar{\ell}} + u_{k\bar{\ell}}) = e^{u+F}\det(g_{k\bar{\ell}}), \\ \text{there exists positive constant } A \text{ such that} \\ \frac{1}{A}(g_{k\bar{\ell}}) \leqslant (g_{k\bar{\ell}} + u_{k\bar{\ell}}) \leqslant A(g_{k\bar{\ell}}). \end{cases}$$

We will follow the same approach as in [C-Y]. One of the basic tools is the maximum principle which can be adapted to our situation.

Maximum principle for functions compatible $\{\pi_i, p_x\}$:

Let U be a C^2 function compatible with $\{\pi_i, p_x\}$ and bounded from above. Then there exists a sequence $\{x_\nu\} \subseteq M\backslash S$ such that:

(i) $\lim_{\nu\to\infty} u(x_\nu) = \sup u$

(ii) $\lim_{\nu\to\infty} |\nabla u(x_\nu)|^2 = 0$

(iii) lim sup of the largest eigenvalue of $(u_{k\bar{\ell}})$ at S_ν with respect to dS^2 is less than or equal to zero.

To solve (1.2) we will use the continuity method as in [C-Y].

Let $\mathcal{C} = \{t \in [0,1] : (2.1)$ can be solved with F replaced by $tF\}$. Clearly $0 \in \mathcal{C}$. The closedness of $\mathcal{C}$ is a consequence of the maximum principle and formulas obtained by local computations and is hence true in our case. The openness of $\mathcal{C}$ needs some discussion. We have to solve the following linear equation:

(1.3) $\Delta v - v = h$,

where $h \in C^{k-2,\delta}(M,\{\pi_i, p_x\})$ and we require that $v \in C^{k,\delta}(M,\{\pi_i, p_x\})$. The key point of solving $\Delta v - v = h$ is that v has to be compatible with $\{\pi_i, p_x\}$. Let $\{M_i\}$ be compact subdomains of $M\backslash S$ with smooth boundary such that $M_i \subseteq M_{i+1}$ and $\bigcup_{i=1}^{\infty} M_i = M\backslash S$. Then let v_i be such that $\Delta v_i - v_i = f$ on M_i and $v_i = 0$ on $\partial M'_i$. First of all by maximum principle v_i is uniformly bounded by the supremum norm of $|f|$. By using interior Schauder estimates we see that a subsequence of v_i converges uniformly on compact subsets of $M\backslash S$ to a solution $\Delta v - v = f$. At this point all we know about v is that v is $C^{k,\delta}$ on $M\backslash S$ and that v is bounded. Then in fact v is compatible with

$\{\pi_i, p_x\}$. The reason is that $\Delta(\pi_i^*(v)) - (\pi_i^*(v)) = \pi_i^*(f)$ or $\Delta(p_x^*(v)) - (p_x^*(v) = p_x^*(f)$ on $\tilde{U}_i \backslash \tilde{A}_i \cup \tilde{D}_i$ or $\tilde{V}_x \backslash \tilde{D}_x$ respectively. $\pi_i^*(f)$ or $p_x^*(f)$ extends as a $C^{k,\delta}$ function on $\tilde{U}_i$ or $\tilde{V}_x$. $\tilde{A}_i, \tilde{D}_i, \tilde{D}_x$ are all of real codimension greater than or equal to two and $\pi_i^*(v)$ or $p_x^*(v)$ are bounded implies that $\pi_i^*(v)$ or $p_x^*(v)$ in fact satisfies the equation on $\tilde{U}_i$ or $\tilde{V}_x$ in the generalized sense. Then standard elliptic regularity theory implies that v is in fact compatible with $\{\pi_i, p_x\}$. This completes the discussion of the openness of $\mathcal{C}$. Also note that the uniqueness of the Monge-Ampère equation also follows from the maximum principle. Now let dS_0^2 denote the Einstein-Kähler metric obtained by solving (1.2). Then we can form the Chern forms $C_1(M, dS_0^2)$ and $C_2(M, dS_0^2)$ from dS_0^2 because they only depend on local formula. As in [Ya1], we have a pointwise inequality.

$$\textbf{(1.4)} \quad (-1)^n C_2(M, dS_0^2) \wedge (C_1(M, dS_o^2))^{n-2} \geqslant \frac{(-1)^n n}{2(n+1)} (C_1(M, dS_0^2))^{n-2}$$

Note that by assumption of compatibility with $\{\pi_i, p_x\}$, the left hand side and right hand side are all integrable and hence we have an inequality by integrating (1.4). Again as in [Ya1] when the integrated inequality becomes an equality we can conclude that (M, dS_0^2) has constant holomorphic sectional curvature.

We will show in the following that if (M, dS_0^2) is of constant holomorphic sectional curvature that $M \backslash C$ is the quotient of the ball. It amounts to construct a simply-connected "covering" of $M \backslash C$ and then show that the "covering" admits a complete metric with constant holomorphic sectional curvature. Then we have to indicate that the local group actions can be patched together.

Let $x_0 \in M\backslash S$. Let γ be a closed curve in $M\backslash C$ based at x_0 then we will say that γ is trivial with respect to $\{\pi_i, p_x\}$ whenever for any tribular neighborhood of γ there is a closed curve γ' in this neighborhood, based at x_0 and homotopic to γ such that: (i) $\gamma' \subseteq M\backslash S$; (ii) γ' is homotopic in $M\backslash S$ to $(\beta_1^{-1} \circ \sigma_1 \circ \beta_1) \circ \cdots \circ (\beta_k^{-1} \circ \sigma_k \circ \beta_k)$ where β_i is a curve in $M\backslash S$ with initial point x_0, σ_i is a closed loop in $M\backslash S$ based at the end-point of β_i and σ_i lies totally inside some V_x or U_i so that its lifting with respect to p_x or π_i is still a closed loop. Then let $\mathcal{C}$ be the collection of all continuous curves $\gamma : [0,1] \to M\backslash C$ such that $\gamma(0) = x_0$. For $\gamma, \gamma' \in$ we say that $\gamma \sim \gamma'$ whenever the closed loop $\gamma'^{-1} \circ \gamma$ is trivial with respect to $\{\pi_i, p_x\}$. Clearly $\sim$ is an equivalence relation. Let $\pi : \mathcal{C}/\sim \to M\backslash C$ be the map which sends an equivalence class of curves to its endpoint. It is routine to check that $\mathcal{C}/\sim$ has a structure of complex manifold and π is a holomorphic map. Moreover, $\pi^*(dS_0^2)$ is a complete metric of constant negative holomorphic sectional curvature. $\mathcal{C}/\sim$ is clearly simply-connected and is hence biholomorphic to the ball B^n. Since B^n is simply-connected and of constant holomorphic sectional curvature we see that local isometries can be extended as global isometries. So $M\backslash C$ is a quotient of the ball.

In the next section we will interpret the meaning of the integral of $C_2(M, dS_0^2) \wedge (C_1(M, dS_0^2))^{n-2}$ and $(C_1(M, dS_0^2))^n$. The following statement summarizes the above discussions.

<u>Theorem 1.2 :</u> Suppose that Condition 1 - Condition 3 are satisfied on M. Then there exists a complete Kähler-Einstein metric dS_0^2 compatible with

$\{\pi_i, p_x\}$. If $\phi : M\backslash S \to M\backslash S$ is a biholomorphic map such that $\pi_i^*(\phi), \pi_i^*(\phi^{-1}), p_x^*(\phi), p_x^*(\phi^{-1})$ can be extended to $\tilde{U}_i$, $\tilde{V}_x$ respectively then dS_0^2 is invariant under ϕ. Let $C_i(M, dS_0^2)$ be the ith Chern form constructed from dS_0^2 then we have:

$$(1.5)\qquad \int_{M\backslash S}(-1)^n\, C_2(M, dS_0^2) \wedge (C_i(M, dS_0^2))^{n-2} \geqslant \frac{(-1)^n n}{2(n+1)} \int (C_1(M, dS_0^2))$$

equality holds implying dS_0^2 has constant negative holomorphic sectional curvature and $M\backslash C$ is a quotient of the ball.

References

[C-G] Carson, J., Griffiths, P., A defect relation for equidimensional holomorphic mappings between algebraic varieties, Ann. Math. **95** (1972).

[C-Y] Cheng, S.Y., Yau, S.T., On the existence of a complete Kähler metirc on non-compact complex manifolds and the regularity of Fefferman's equation, Comm. Pure Appl. Math. **33** (1980), 507-544.

[Hi] Hirzebruch, F., Hilbert modular surfaces, L'Enseignment Mathemétique **19** (1973), 183-282.

[Mi1] Miyaoka, Y., On the Chern numbers of surfaces of general type, **42** (1977), 225-237.

[Mi2] Miyaoka, Y., The maximal number of quotien singularities on surfaces with given numerical invariants, Math. Ann. **268** (1984), 159-171.

[Mu] Mumford, D., Hirzebruch's proportionality theorem in the non-compact case, Inv. Math. **42** (1977), 239-272.

[Ya1] Yau, S.T., Calabi's conjecture and some new results in algebraic geometry, Proc. Natl. Acad. Sci. USA **74**, No. 5, May 1977.

[Ya2] Yau, S.-T., Métriques de Kähler-Einstein sur les variétés ouvertes, in Premiere Classe de Chern et Courbure de Ricci : Preuve de la Conjecture de Calabi, Séminaire Palaiseau, France (1978), 163-167.

[Sa] Sakai, F., Semi-stable curves on algebraic surfaces and logarithmic pluricanonical maps, Math. Ann. **254** (1980), 89-120.

Contemporary Mathematics
Volume 49, 1986

LAPLACIAN ON MANIFOLDS AND ANALOGOUS DIFFERENCE OPERATOR FOR GRAPHS

Jozef Dodziuk[1]

ABSTRACT. Rectangular lattice in R^n is often thought of as a discrete model of Euclidean space. An arbitrary graph might be considered as a discrete analog of a space of variable curvature. We explore some analogies (e.g. Cheeger's bound for the bottom of the spectrum) and some differences between the Laplacian on manifolds and its discrete analog.

In their beautiful paper [3] Courant, Friedrichs and Lewy exploited the analogies between the Laplacian Δ in the plane and its discrete analog Δ_h

$$(1)\qquad \Delta_h u(ih,\ jh) = (1/h^2)(u((i+1)h,\ jh) + u(ih,\ (j+1)h) + u((i-1)h,\ jh) + u(ih,\ (j-1)h) - 4u(ih,\ jh)),$$

on the lattice hZ^2. They observed that Δ_h has many properties (the maximum principle, the relation with random walk) analogous to properties of Δ and used these formal analogies to prove that solutions of suitably formulated Dirichlet problem for Δ_h converged to solutions of Dirichlet problem for Δ as $h \to 0$. The operator Δ_h continues to play a very important role in mathematical physics. The viewpoint taken by physicists is that the lattice hZ^2 is a good approximation of R^2, hence the study of Δ_h yields information about Δ. Clearly, one is not limited to two dimensions here.

What if the universe is not a Euclidean space? There are standard methods of manufacturing a graph from a given manifold M (e.g. take a one dimensional skeleton of a triangulation of M). My purpose is to show that, given any graph, there is a natural difference operator with properties remarkably similar to properties of the Laplacian on a Riemannian manifolds. I will describe only formal analogies here. However, in favorable circumstances, they can be used to obtain information about the Laplacian on manifolds by studying its discrete analog.

From now on K will denote a graph, i.e. a one dimensional simplicial complex. Consider the simple random walk on K in which the probability

---

1980 Mathematics Subject Classification Primary 39A10; Secondary 60J15, 58G99.

[1] Supported in part by NSF Grant No. MCS8301890

$p(x,y)$ of moving from a vertex x to a neighboring vertex y is equal to m_x^{-1}, where m_x is the number of edges meeting at x. Successive moves are assumed to be independent. It is very easy to see that the probabilities $p_n(x,y)$ of going from x to y in n steps satisfy the equation

$$p_{n+1}(x,y) - p_n(x,y) = \sum_{z\sim x} p(x,z)p_n(z,y) - p_n(x,y) = m_x^{-1} \sum_{z\sim x} p_n(z,y) - p_n(x,y) ,$$

where $z\sim x$ means that z is joined to x by an edge. If we set $u(x,n) = p_n(x,y)$, then the equation above becomes

$$u(x,n+1) - u(x,n) = m_x^{-1} \sum_{z\sim x} u(z,n) - u(x,n).$$

This is the difference analog of the heat equation $\frac{\partial u}{\partial t} = \Delta u$, and the left-hand side becomes a natural candidate for the difference analog Δ^c of Δ.

(2) $$\Delta^c u(x) = m_x^{-1} \sum_{z\sim x} u(z) - u(x).$$

Note that (1) is a special case.

The same operator arises in another context. Suppose that K is a one dimensional skeleton of a higher dimensional finite simplicial complex L. Since L is finite we can identify real-valued cochains with chains. Let d and ∂ denote the boundary and coboundary operations.

$$0 \longrightarrow C^0(L) \underset{\partial}{\overset{d}{\rightleftarrows}} C^1(L) \underset{\partial}{\overset{d}{\rightleftarrows}} C^2(L) \;\ldots\ldots .$$

Then Δ^c defined above coincides with the restriction of $-(d\partial + \partial d)$ to 0-cochains. The operator $-(d\partial + \partial d)$ is an analog of the Hodge Laplacian on forms. Together with closely related operators it was used to "approximate" the Hodge theory (cf. [4], [5], [6], [11]).

We now return to our random walk and the Laplacian Δ^c for a graph K. It is an immediate consequence of (2) that, if $\Delta^c u \geq 0$ and u attains a local maximum at $x \in K$, $u(x) = u(y)$ for $y\sim x$. Similarly if $\Delta^c u \leq 0$ and $u \geq 0$ then $m_x^{-1}u(y) \leq u(x) \leq m_y u(y)$. These two facts are analogs of the maximum principle and the Harnack inequality respectively. They are used in the proof of Theorem 1 below. To state this result we need some notation. Let $\ell^2(K) = \{u | \Sigma_{x\in K} |u(x)|^2 < \infty\}$. This is a Hilbert space with an obvious inner product and Δ^c is a (possibly unbounded) operator on $\ell^2(K)$. We define $\lambda_o(K)$ to be the infimum of the spectrum of $-\Delta^c$. Clearly

$$\lambda_o(K) - \inf \frac{(-\Delta^c u,u)}{(u,u)} ,$$

where the infimum is taken over all functions with finite support. The following theorem proved in [5] is a discrete version of Theorem 1 of [8].

THEOREM 1. *Let $\lambda \leq \lambda_o(K)$. The equation $\Delta^c u + \lambda u = 0$ has a positive solution.*

We now define an analog of Cheeger's constant of an infinite graph. For a finite subgraph L K, define #L to be the number of vertices of L, and #∂L to be the number of vertices of L with at least one neighbor in the complement of L. We then set

$$h(K) = \inf \frac{\#\partial L}{\#L} ,$$

with the infimum taken over all finite subgraphs of K.

THEOREM 2. *Suppose there exists an integer m such that $m_x \leq m$ for all $x \in K$. Then, if $h = h(K)$,*

$$c_1 h^2 \leq \lambda_o(K) \leq c_2 h ,$$

with the constants c_1 0, c_2 0 depending only on m.

The first inequality is an analog of Cheeger's [2] lower bound for the bottom of the spectrum of the Laplacian on a Riemannian manifold, and the second one is the discrete version of the result of Buser [1]. The proof can be found in [5].

Another striking parallel between the theories of Δ and Δ^c is the criterion for transience (of either random walk or the Brownian motion) due in its definitive form to Lyons [9] in the discrete case and to Lyons and Sullivan [10] in the smooth setting. It has quite a long history (reviewed in [LS]) in which the "discrete" is often not separated from the "continuous" (cf. Theorem 11 of [12]). Recall that transience means that the randomly moving particle has positive probability of escaping to infinity. This is equivalent to existence of Green's function for the Laplacian, which in turn happens if and only if a nonconstant positive superharmonic function exists.

THEOREM 3. *Let K be an infinite graph. The random walk on K is transient if and only if there exists a one-cochain X on K such that*

(a) $\quad \Sigma < X, e>^2 < \infty$,

with the summation over all edges e of K and <X, e> denoting the value of X on e.

(b) $\quad \sum_x |\partial X(x)| < \infty$,

with the summation over all vertices of K.

(c) $\quad \sum_x \partial X(x) > 0$.

In the smooth setting one replaces K by a manifold, X by a vector field. The conditions (a), (b), (c) are replaced by $\int|X|^2 < \infty$, $\int|\mathrm{div}X| < \infty$, $\int\mathrm{div}X > 0$ respectively.

In view of the analogies discussed above one may hope that some open problems for the Laplacian on manifolds can be attacked by first considering related questions in the discrete setting. I believe that this helps in focusing on the main issues. However, it has to be mentioned that the analogy is not perfect. A phenomenom which may occur in the smooth setting, but is impossible for a graph is explosion. This happens, by definition, if a Brownian particle has nonzero probability of reaching infinity in finite time. The difference analog of the heat equation has finite propagation speed, which explains why explosion can not occur for a random walk on a graph. Another criticism is that the Laplacian on a manifold depends on the metric which can be varied. Our formula (2) for Δ^c has no parameters which would allow changing the operator. I will describe now very briefly a generalization of simple random walks, which recently found rather deep applications to problems concerning the Laplacian on open manifolds. Think of the graph K as an electrical network. The simple random walk described above corresponds to having one ohm resistor along every edge. One can vary the resistances and this corresponds to changing the Riemannian metric of a manifold. If $R_{xy} = C_{xy}^{-1}$ is the resistance between two vertices x and y, one defines the transition probabilities

$$p(x,y) = (\sum_{z\sim x} C_{xz})^{-1} C_{xy} .$$

One can define Δ^c now, generalizing (2), as follows.

$$\Delta^c u(x) = \sum_{z\sim x} p(x,z)u(z) - u(x) .$$

A beautiful elementary discussion of these ideas can be found in [7]. Quite recently P. Doyle and T. Lyons (unpublished) independently gave examples of two quasi-isometric Riemannian surfaces one of which admits nonconstant positive harmonic functions, while the other does not. In both cases the starting point is a construction of an "electrical network" as above for which one can eliminate or create harmonic functions by small changes in resistances. Thinking of edges of the graph as tubes of uniform diameter one then creates a surface by taking tubes of lengths proportional to resistances of corresponding edges and connecting them at nodes of the graph. It can be proved that the behavior of harmonic functions for the resulting surface is the same as it was for the "electrical network" .

It is very likely that this point of view will find many applications in the future.

BIBLIOGRAPHY

1. P. Buser, "A note on the isoperimetric constant," Ann. Scient.Ec.Norm. Sup. 15(1982), 213-230.

2. J. Cheeger, "A Lower bound for the smallest eigenvalue of the Laplacian," in Problems in Analysis, a symposium in honor of S.Bochner, Princeton University Press, Princeton, 1970, 195-199.

3. R. Courant, K. Friedrichs, and H. Lewy, "Uber die partiellen Differenzengleichungen der mathematischen Physik," Math. Ann. 100 (1928), 32-74. Translation: On the Partial Difference Equations of Mathematical Physics, IBM J. Research and Development, 11(1967), 215-237.

4. J. Dodziuk, "Finite-difference approach to the Hodge theory of harmonic forms," Amer. J. Math. 98(1976), 79-104.

5. J. Dodziuk, "Difference equations, isoperimetric inequality and transience of certain random walks," Trans. Amer. Math. Soc. 294(1984), 787-794.

6. J. Dodziuk and V.K. Patodi, "Riemannian structures and triangulations of manifolds," J. Indian Math. Soc. 40(1976), 1-52.

7. P.G. Doyle and J.L. Snell, Random walks and electric networks, MAA, 1984.

8. D. Fischer-Colbrie and R. Schoen, "The structure of complete stable minimal surfaces in 3-manifolds of non-negative scalar curvature," Comm. Pure Appl. Math. 33(1980), 199-211.

9. T. Lyons, "A simple criterion for transience of a reversible Markov Chain," Ann. of Probability ii(1983), 393-402.

10. T. Lyons and D. Sullivan, "Function Theory, random paths and covering spaces," J. Differential Geometry 19(1984), 299-323.

11. W. Müller, "Analytic torsion and R-torsion of Riemannian manifolds," Advances in Math. 28(1978), 233-305.

12. H.L. Royden, "Harmonic functions on open Riemann surfaces," Trans. Amer. Math. Soc. 73(1952), 40-94.

DEPARTMENT OF MATHEMATICS
QUEENS COLLEGE OF CUNY
FLUSHING, N.Y. 11367

Contemporary Mathematics
Volume **49**, 1986

ON ISOTROPIC HARMONIC MAPS TO REAL AND QUATERNIONIC GRASSMANNIANS

J. F. Glazebrook

ABSTRACT. For any Riemann surface M, the Erdem-Wood theorem which classifies full, isotropic harmonic maps $\phi:M \to G_k(\mathbb{C}^N)$, can be restricted to classify such maps to the Grassmannian $G_k(\mathbb{R}^N)$ of real k-planes in $\mathbb{R}^N$. The author has shown how analogous observations for this latter case can be exploited to classify certain harmonic maps from M to n-dimensional quaternionic projective space $\mathbb{H}P^n$. Together with including the details for the case of the quaternionic Grassmannian $G_k(H^N)$, we present some further results for the real (oriented and non-oriented) Grassmanians. In particular, we establish what is essentially a 2:1 correspondence between full, isotropic harmonic maps $\phi:M \to G_k^0(\mathbb{R}^{2r+k})$ (real oriented k-planes in $\mathbb{R}^{2r+k}$) and certain holomorphic subbundles V of rank r of the trivial complex $(2r+k)$ - bundle $\underline{\mathbb{C}}^{2r+k}$ on M.

1. THE ERDEM-WOOD THEOREM

We refer to the excellent report [4] for properties relating to harmonic maps. In this section we shall briefly recall the terminology and the main result of [6], the notation of which we shall use freely.

Throughout, M will be taken to denote a Riemann surface (open or closed), and V and X will denote holomorphic subbundles of the trivial N-plane bundle, $\underline{\mathbb{C}}^N$, over M.

Let $L \to G_k(\mathbb{C}^N)$ be the tautological k-plane bundle. Given a smooth map $\phi:M \to G_k(\mathbb{C}^N)$, we may take its universal lift Φ, to be a section of the bundle $\mathrm{Hom}\,(\phi^{-1}L,\underline{\mathbb{C}}^N)$ on M. If D denotes covariant differentiation in this bundle, then D has a splitting into (1,0) and (0,1) types denoted by D' and D'' respectively. We may consider the iterated derivatives, $D'^{\alpha}\Phi$ and $D''^{\beta}\Phi$ of all orders $\alpha\ \beta \geq 0$; these are systematically derived from [6, Lemma 3.1].

Let U be an open set in M; then taking the smooth section

1980 Mathematics Subject Classification 58E20.

$W \in C_U(\mathrm{Hom}\,\phi^{-1}L, \underline{\mathbb{C}}^N))$, we shall denote by $\mathrm{Im}(W_x)$, the image of the linear map

(1.1) $\quad W_x : L_{\phi(x)} \to \mathbb{C}^N$.

On setting

(1.2) $\quad \mathrm{Im}\, W = \bigcup_{x \in U} (\{x\} \times \mathrm{Im}(W_x)) \subset U \times \mathbb{C}^N$,

we say that a smooth map $\phi : M \to G_k(\mathbb{C}^N)$ is <u>harmonic</u> if, on each chart U ,

(1.3) $\quad \mathrm{Im}(D''D'\Phi) \subset \mathrm{Im}(\Phi)$.

The map is said to be <u>isotropic</u> if

(1.4) $\quad \mathrm{Im}(D'^{\alpha}\Phi) \perp \mathrm{Im}(D''^{\beta}\Phi)$, for all $\alpha, \beta \geq 0$, $\alpha + \beta \geq 1$.

For a smooth map $\phi : M \to G_k(\mathbb{C}^N)$, we have the (unique) <u>associated subbundles</u>, $\phi'_{(\alpha)}$, $\phi''_{(\alpha)}$ of $\underline{\mathbb{C}}^N$, for each $\alpha \geq 1$, such that for all $x \in M$

(1.5) $\quad (\phi'_{(\alpha)})_x = \mathrm{span}\{\mathrm{Im}(D'^{\gamma}\Phi)_x : 1 \leq \gamma \leq \alpha\}$

(1.6) $\quad (\phi''_{(\beta)})_x = \mathrm{span}\{\mathrm{Im}(D''^{\delta}\Phi)_x : 1 \leq \delta \leq \beta\}$

where the span takes its maximum dimension in each case. This span is independent of the chart chosen for the definition. The <u>augmented associated bundle</u> of $\phi'_{(\infty)}(\phi''_{(\infty)})$ is defined to be

(1.7) $\quad \tilde{\phi}'_{(\infty)} = \phi'_{(\infty)} + \phi$ (respectively, $\tilde{\phi}''_{(\infty)} = \phi''_{(\infty)} + \phi$) .

DEFINITION 1.1. Let V and X be defined as above. Then we say that (V,X) is a ∂'-<u>pair of rank difference</u> k if:

(a) $V \subset X$ (b) $\mathrm{rank}(X) - \mathrm{rank}(V) = k$ (c) $\partial'C(V) \subset C(X)$.

Let A be a subspace of $\mathbb{C}^N$, then a subbundle of the form $M \times A$ is called a <u>constant subbundle</u> of $\mathbb{C}^N$.

DEFINITION 1.2. A ∂'-pair (V,X) of $\underline{\mathbb{C}}^N$ is <u>full</u> if

a) the only constant subbundle of $\underline{\mathbb{C}}^N$ containing X is $\underline{\mathbb{C}}^N$ itself;

b) the only constant subbundle of $\underline{\mathbb{C}}^N$ contained in V is the zero subbundle $\underline{0} = M \times \{0\}$.

We say that a map $\phi : M \to G_k(\mathbb{C}^N)$ is <u>full</u> if the only subspace of $\underline{\mathbb{C}}^N$ containing each subspace $\phi(x)$, for $x \in M$, is $\mathbb{C}^N$ itself.

THEOREM 1.3. [6,Th.1.1]. <u>There exists a bijective correspondence between full -pairs</u> (V,X) <u>of rank difference</u> k, <u>on</u> M, <u>and full, isotropic harmonic</u>

maps $\phi : M \to G_k(\mathbb{C}^N)$ *given by*

(1.8) $\phi(x) = V_x^{\perp} \cap X_x$

for all $x \in M$.

Conversely, given such a ϕ , *the pair* $(\phi''_{(\infty)}, \tilde{\phi}''_{(\infty)})$ *is a full* ∂'*-pair.*

2. REAL GRASSMANNIANS

In this section we consider the Grassmannian $G_k(\mathbb{R}^N)$ of (non-oriented) real k-planes in $\mathbb{R}^N$. When the '*total isotropy*' condition

(2.1) $\bar{X} = V^{\perp}$

is imposed on the ∂'-pair (V,X) above then Theorem **1.3.** restricts to classify full, isotropic harmonic maps $\phi : M \to G_k(\mathbb{R}^N)$ [6, Cor. 1.8]. Here, the fullnes and isotropy of the map is that into $G_k(\mathbb{C}^N)$ on taking the totally geodesic inclusion.

We aim to re-state the above result in terms of V alone. Firstly, we establish a lemma:

LEMMA 2.1. *Let* (V,X) *be a* ∂'*-pair of holomorphic subbundles of* $\mathbb{C}^N$ *on* M . *Assume that* V *and* X *satisfy condition* (2.1); *then* V *satisfies the conditions*

$$(2.2) \quad \begin{cases} \bar{V} \subset V^{\perp} \\ \partial' C(V) \subset C(V^{\perp}). \end{cases}$$

Proof. This is an immediate consequence of conditions a) and c) in Definition 1.1 and the condition (2.1).

Let $(\, , \,)^{\mathbb{C}}$ denote the usual $\mathbb{C}$-bilinear symmetric inner product on $\mathbb{C}^N$. Since $(v,w)^{\mathbb{C}} = \langle v,\bar{w} \rangle$ (the usual Hermitian inner product) for $v,w \in \mathbb{C}^N$, the inclusion $\bar{V} \subset V^{\perp}$ is equivalent to V being isotropic for $(\, , \,)^{\mathbb{C}}$, i.e. $(V,V)^{\mathbb{C}} = 0$. In order to make the above conditions more explicit, we note that since V is a holomorphic subbundle of $\underline{\mathbb{C}}^N$ of rank t say, on M, it therefore corresponds (bijectively) to a holomorphic map $\psi : M \to G_t(\mathbb{C}^N)$. The conditions in (2.2) may be interpreted as $(\psi(x),\psi(x))^{\mathbb{C}} = 0$ and $(\psi(x),\partial'\psi(x))^{\mathbb{C}} = 0$, respectively, for $x \in M$.

Given a V satisfying the conditions in (2.2), we seek an extra condition to ensure that the resulting ∂'-pair (V,X) related by (2.1), is full in the sense of Definition 1.2. The condition is that V should satisfy condition ii) in Definition 1.2, namely, the only constant subbundle of $\mathbb{C}^N$ contained in V , is the zero subbundle $\underline{0} = M \times \{0\}$. Since $\bar{X} = V^{\perp} \subset M \times \mathbb{C}^N$, it therefore follows that the only constant subbundle of $\underline{\mathbb{C}}^N$ containing X, is $\underline{\mathbb{C}}^N$ itself. If V satisfies this condition, then we shall say that V is <u>full</u>.

Observe, that apart from the case k and $N-k$ both odd, any Grassmannian $G_k(\mathbb{R}^N)$ may be expressed as $G_k(R^{2r+k}), (\cong G_{2r}(\mathbb{R}^{2r+k}))$ for some $r \geq 1$. We shall restrict our attention to $G_k(\mathbb{R}^{2r+k})$ as a consequence of the following:

PROPOSITION 2.2. *For p and q both odd, there exist no full isotropic harmonic maps $\phi: M \to G_p(\mathbb{R}^{p+q}) (\cong G_q(\mathbb{R}^{p+q}))$.*

Proof. Assume on the contrary that such a map did exist. Then by imposing condition (2.1) on the corresponding ∂'-pair (V,X) obtained by virtue of Theorem 1.3, we would have

$$\text{Rank } X = \text{Rank } \bar{X} = (p+q) - \text{Rank } V$$

and $\text{Rank } X - \text{Rank } V = p$. But for p and q both odd, this would lead to impossible data.

Remark. The case $p = 1$, q odd was settled by Calabi [2] (see also [5]).

Thus restricting our attention to $G_k(\mathbb{R}^{2r+k})$, and taking conditions (2.1) and (2.2) into account, we see that in this case, the correct choice of ranks for V and X will be:

$$(2.3) \qquad \begin{cases} \text{Rank } V = r \\ \text{Rank } X = r+k \\ (\text{Rank } X^{\perp} = r) . \end{cases}$$

Thus we can state:

COROLLARY 2.3 (See also [6, Cor. 1.8]). *There exists a bijective correspondence between full, isotropic harmonic maps $\phi: M \to G_k(\mathbb{R}^{2r+k})$ and full, holomorphic subbundles V of $\underline{\mathbb{C}}^{2r+k}$ of rank r on M, satisfying the conditions (2.2).*

Proof. The conditions satisfied by V, together with the relationship (2.1), imply the existence of a holomorphic subbundle X of $\underline{\mathbb{C}}^{2r+k}$ of rank $r+k$ on M, such that (V,X) is a full ∂'-pair of rank difference k.

The map ϕ is defined by (1.8), and in the inverse assignment, V is set equal to $\phi''_{(\infty)}$, once we have taken the (totally-geodesic) inclusion into $G_k(\mathbb{C}^{2r+k})$.

EXAMPLE 2.4. Here is an example for $r = 1$ and $k = 3$. We take V to be the full holomorphic map $f:\mathbb{C}P^1 \to \mathbb{C}P^4$ defined as follows [3]:
For $z \in \mathbb{C}P^1$, let $f(z) = (\xi_1,\ldots,\xi_5)$ where

$$\xi_1 = 2\sqrt{3}(1+z^4) \qquad \xi_3 = 2\sqrt{3}i(1-z^4) \qquad \xi_5 = \sqrt{3}(-4z+4z^3)$$
$$\xi_2 = 12z^2 \qquad \xi_4 = \sqrt{3}i(-4z-4z^3) .$$

Regarded as conditions on the map f, both conditions in (2.2) are satisfied, as a straightforward calculation shows. We obtain a ∂'-pair (V,X) of rank difference 3, by setting $X = \bar{f}^{\perp} = f_3$, <u>the</u> 3rd <u>associated curve of</u> f [5] [11], and X has rank 4 as a holomorphic subbundle of $\underline{\mathbb{C}}^5$. The pair (V,X) satisfies the total isotropy condition (2.1), and from (1.8), the map $\phi:S^2 \to G_3(\mathbb{R}^5)$ defined by $\phi(x) = f^{\perp}(x) \cap f_3(x)$, is a full isotropic harmonic map.

3. THE TWISTOR SPACE OF $G_k(\mathbb{R}^{2r+k})$

Consider the flag manifold

$$F^{2r+k}_{r,r} = \{(P,Q) \in G_r(\mathbb{C}^{2r+k}) \times G_r(\mathbb{C}^{2r+k}) : P \perp Q\}$$
$$\cong U(2r+k)/U(r) \times U(r) \times U(k) .$$

We obtain a fibration

$$(3.1) \qquad \pi:F^{2r+k}_{r,r} \longrightarrow G_k(\mathbb{C}^{2r+k}) (\cong U(2r+k)/U(2r) \times U(k))$$

induced by the inclusion of $U(r) \times U(r)$ in $U(2r)$ in the standard way. For the natural homogeneous space metrics on $F^{2r+k}_{r,r}$ and $G_k(\mathbb{C}^{2r+k})$, π is a Riemannian submersion [5].

Given a full ∂'-pair (V,X) of $\underline{\mathbb{C}}^{2r+k}$ satisfying the rank conditions in (2.3), we may define a harmonic map $\psi:M \to F^{2r+k}_{r,r}$ by $\psi(x) = (V_x, X_x^{\perp})$, for $x \in M$. This map may be seen to be π-horizontal to the fibration (3.1). Now if condition (2.1) is satisfied (and hence (2.2)), we see that V_x and $X_x^{\perp}$ are both r-<u>dimensional isotropic subspaces of</u> $\mathbb{C}^{2r+k}$ <u>with respect to</u> $(\, , \,)^{\mathbb{C}}$. Their totality is a certain quadric Grassmannian which we shall denote by $S_{k,r} \cong O(2r+k)/U(r) \times O(k)$ [9] (see also [1] [10]). In particular $S_{1,r} \cong H_r$, is the twistor space for $\mathbb{R}P^{2r}$ (and S^{2r}) as discussed in [5].

The space $S_{k,r}$ is a totally geodesic submanifold of $F^{2r+k}_{r,r}$, via the totally geodesic inclusion $P \to (P,\bar{P})$, where $P \in S_{k,r}$, and $S_{k,r}$ will be the receiving space for the map ψ as defined above. Retaining π to denote the restriction of (3.1) to $S_{k,r}$, we obtain the fibration

$$(3.2) \qquad \pi: S_{k,r} \longrightarrow G_k(\mathbb{R}^{2r+k})$$

which as a homogeneous fibration is represented by

$$(3.3) \qquad \pi: O(2r+k)/U(r) \times O(k) \to O(2r+k)/O(2r) \times O(k)$$

with fibre $O(2r)/U(r)$. The inclusion associated with (3.1) induces the inclusions $U(r) \hookrightarrow O(2r) \hookrightarrow O(2r) \times O(k)$.

Because of the conditions imposed on V (and $X^{\perp}$), the map $\psi: M \to S_{k,r} \subset F^{2r+k}_{r,r}$ may be defined more simply by $\psi(x) = V_x$. Again, ψ is π-horizontal to (3.2) (essentially as a result of conditions (2.2)), and using the fact that (3.2) is a Riemannian submersion, the map $\phi: M \to G_r(\mathbb{R}^{2r+k})$ defined by $\phi(x) = \pi \circ V_x$, is harmonic, following a standard composition principle [4]. Furthermore, assuming V is full, ϕ is a full isotropic harmonic map. This construction is intrinsically contained in the proof of Theorem 1.3 and its restriction to $G_k(\mathbb{R}^{2r+k})$ as expressed by Corollary 2.3.

4. REAL ORIENTED GRASSMANNIANS

The Grassmannian $G^0_k(\mathbb{R}^{2r+k})$ of oriented real k-planes in $\mathbb{R}^{2r+k}$ is a simply connected Riemannian symmetric space with coset space representation $SO(2r+k)/SO(2r) \times SO(k)$. It is the orientable double covering of $G_k(\mathbb{R}^{2r+k})$ for which the covering projection (that forgets orientation)

$$(4.1) \qquad G^0_k(\mathbb{R}^{2r+k}) \longrightarrow G_k(\mathbb{R}^{2r+k}),$$

is a local isometry.

LEMMA 4.1. _Let_ $\phi: M \to G_k(\mathbb{R}^{2r+k})$ _be an isotropic harmonic map. Then_ ϕ _lifts as a harmonic map to_ $G^0_k(\mathbb{R}^{2r+k})$.

Proof. Such a map ϕ has an associated lift $\psi: M \to S_{k,r}$ that is harmonic and π-horizontal with respect to (3.2). For $W = \psi(x) \in S_{k,r}$, we can choose an oriented orthonormal basis $\{e_j\}$, $j = 1,\dots,2r+k$, such that W is spanned by $\{e_1 + ie_2, \dots\dots, e_{2r-1} + ie_{2r}\}(i = \sqrt{-1})$. This choice of an oriented basis sets W in the space $S^0_{k,r} \cong SO(2r+k)/U(r) \times SO(k)$, and we have a fibration

$$(4.2) \qquad \tilde{\pi}: S^0_{k,r} \cong SO(2r+k)/U(r) \times SO(k) \to G^0_k(\mathbb{R}^{2r+k}) \cong SO(2r+k)/SO(2r) \times SO(k)$$

induced by the inclusions $U(r) \hookrightarrow SO(2r) \hookrightarrow SO(2r) \times SO(k)$; explicitly $\tilde{\pi}(W) = e_{2r+1} \wedge \ldots \wedge e_{2r+k}$.

The map ψ remains harmonic and is $\tilde{\pi}$ -horizontal with respect to (4.2). Applying the previous composition principle, the map $\tilde{\phi} = \tilde{\pi} \circ \psi$ is a harmonic map from M to $G_k^0(\mathbb{R}^{2r+k})$.

By choosing a basis with the opposite orientation to that of the first, we may repeat this procedure to produce another lift of ϕ .

The two resulting maps ϕ_+ and ϕ_- , say, cover the map ϕ to $G_k(\mathbb{R}^{2r+k})$.

We shall say that a map $\tilde{\phi}:M \to G_k^0(\mathbb{R}^{2r+k})$ is <u>isotropic</u> if $\phi:M \to G_k(\mathbb{R}^{2r+k})$ is isotropic where we take Φ to be the lift of ϕ. Similarly, we assign the notion of fullness of $\tilde{\Phi}$ when ϕ itself is full. Thus combining Corollary 2.3 and Lemma 4.1, we obtain:

COROLLARY 4.2. <u>There exists a 1:1 correspondence between pairs</u> (ϕ_+,ϕ_-) <u>of full, isotropic harmonic maps</u> $M \to G_k^0(\mathbb{R}^{2r+k})$ <u>and full holomorphic subbundles</u> V <u>of</u> $\mathbb{C}^{2r+k}$ <u>of rank</u> r <u>on</u> M , <u>satisfying the conditions</u> (2.2).

For $2r = n$ and $k = 2$, the above corollary classifies full, isotropic harmonic maps $M \to Q_n \cong G_2^0(\mathbb{R}^{n+2})$, the complex quadric in $\mathbb{C}P^{n+1}$. For n odd, we must take the equivalent representation $G_n^0(\mathbb{R}^{n+2})$, i.e. set $k = n$ and $r = 1$.

5. QUATERNIONIC GRASSMANNIANS

Let $\mathbb{H}$ denote the division ring of quaternions. For $q \in \mathbb{H}$, we write $q = a + bj$ where $a,b \in \mathbb{C}$ and j is a unit quaternion. We thus obtain the identification $\mathbb{H} \cong \mathbb{C} + \mathbb{C}j \cong \mathbb{C}^2$, and this can be generalised to N quaternionic dimensions to obtain the identification $\mathbb{H}^N \cong \mathbb{C}^{2N}$. Let

(5.1) $\sigma : \mathbb{C}^{2N} \to \mathbb{C}^{2N}$

be the conjugate linear map induced by left multiplication by j. The Grassmannian $G_k(\mathbb{H}^N)$ ($\cong Sp(N)/Sp(k) \times Sp(N-k)$ of quaternionic k-dimensional. planes in $\mathbb{H}^N$ may be viewed as the Grassmannian of complex 2k-planes in $\mathbb{C}^{2N}$ that are 'real' or fixed under the action of σ . This is a straightforward generalisation of the underlying principle for obtaining the canonical twistor fibration

(5.2) $\mathbb{C}P^{2n+1} \to \mathbb{H}P^n$

with fibre $\mathbb{C}P^1$.

Given a full ∂'-pair (V,X) of $\underline{\mathbb{C}}^{2N}$ of rank difference $2k$, we can construct a full, isotropic harmonic map $M \to G_{2k}(\mathbb{C}^{2N})$ by (1.8). When the condition

$$(5.3) \qquad \sigma X = V^{\perp} \quad \text{(or equivalently, } \sigma V = X^{\perp})$$

is satisfied, then the map ϕ has image in $G_k(\mathbb{H}^N)$.

We may define a non-degenerate anti-symmetric inner product S for vectors in $\mathbb{C}^{2N} \cong \mathbb{H}^N$, in terms of the usual Hermitian inner product $<,>$ on $\mathbb{C}^{2N}$, by

$$(5.4) \qquad S(x,y) = <x,\sigma y> \quad \text{for } x,y \in \mathbb{C}^{2N}.$$

In terms of V alone, the analogous consitions to (2.2) are seen to be

$$(5.5) \quad \begin{cases} V \subset V^{\perp_S} \\ \sigma\partial' C(V) \subset C(V^{\perp}) \end{cases}$$

where $\perp_S$ denotes orthogonality with respect to S. These conditions can be explicated analogously to those in (2.2).

For this case, the correct choice of ranks for V and X is

$$(5.6) \quad \begin{cases} \text{Rank } V = N-k \\ \text{Rank } X = N+k \\ (\text{Rank } X^{\perp} = N-k) \end{cases}$$

Now $G_k(\mathbb{H}^N)$ may be seen to be totally geodesic in $G_{2k}(\mathbb{C}^{2N})$, and as was the case for the maps into $G_k(\mathbb{R}^N)$, we shall say that a map $\phi: M \to G_k(\mathbb{H}^N)$ is <u>isotropic</u> if the resulting map into $G_{2k}(\mathbb{C}^{2N})$ is isotropic, on taking the totally geodesic inclusion. Similarly, the notion of fullness is assigned. Thus a straighforward generalisation of the main result of [8] is:

COROLLARY 5.1. <u>There exists a bijective correspondence between full, isotropic harmonic maps</u> $\phi: M \to G_k(\mathbb{H}^N)$ <u>and full, holomorphic subbundles</u> V <u>of</u> $\underline{\mathbb{C}}^{2N}$ <u>of rank</u> $N-k$ <u>on</u> M, <u>satifying conditions</u> (5.5).

For $x \in M$, we see that V_x and $X_x^{\perp}$ are both $(N-k)$-dimensional S-isotropic subspaces of $\mathbb{C}^{2N}$. Their totality is the quadric Grassmannian $\mathcal{R}_{k,N-k} \cong Sp(N)/Sp(k) \times U(N-k)$ ([1], [8], [9], [10]). The space $\mathcal{R}_{k,N-k}$ is totally geodesic in the flag manifold $F^{2N}_{N-k,N-k} \cong U(2N)/U(N-k) \times U(N-k) \times U(2k)$, and the fibration $\mathcal{F}^{2N}_{N-k,N-k} \longrightarrow G_{2k}(\mathbb{C}^{2N})$ restricts to give a fibration

(5.6) $\pi:\mathcal{R}_{k,N-k} \cong Sp(N)/Sp(k) \times U(N-k) \to G_k(\mathbb{H}^N) \cong Sp(N)/Sp(k) \times Sp(N-k)$

given by the inclusion of $U(N-k)$ in $Sp(N-k)$, with fibre $Sp(N-k)/U(N-k)$.

The map $\phi:M \to G_k(\mathbb{H}^N)$ defined by

(5.7) $\phi(x) = \pi \circ V_x = (V_x + \sigma V_x)^{\perp}$

for $x \in M$, may be seen to be the construction of the full, isotropic harmonic map given by Corollary 5.1.

Note that for $k = n$, $N = n+1$, $\mathcal{R}_{n,1} \cong \mathbb{C}P^{2n+1}$ and for $k = 1$, $N = n+1$, $\mathcal{R}_{1,n} \cong \mathcal{T}_n$ as discussed in [8]. As twistor spaces, $\mathcal{R}_{n,1}$ and $\mathcal{R}_{1,n}$ correspond to two particular cases of Corollary 5.1 for isotropic harmonic maps into $\mathbb{H}P^n$. These are maps for which Rank $\phi''_{(\infty)} = 1$ and Rank $\phi''_{(\infty)} = n$, respectively.

Remark. As in the case of $G_k(\mathbb{R}^{2r+k})$, we see that conditions (5.5) determine the π-horizontally of the holomorphic map $\psi:M \to \mathcal{R}_{k,N-k}$ determined by the holomorphic subbundle V.

6. SOME EXAMPLES USING REPRESENTATION THEORY

EXAMPLE 6.1. This is a generalisation of an example in [10] to which we refer for terminology and further details. We shall consider the case $G_k(\mathbb{H}^{n+k})$ and proceed to consider the fibration

(6.1) $\pi:Z \cong Sp(n+k)/Sp(k) \times U(n) \to G_k(\mathbb{H}^{n+k}) = Sp(n+k)/Sp(k) \times Sp(n)$

where we take Z to denote $\mathcal{R}_{k,n}$ for simplicity.

Let $E \cong \mathbb{C}^{2n}$, $G \cong \mathbb{C}^{2K}$ and $K \cong \mathbb{C}^n$ be the respective basic representations of the complexified subgroups $Sp(n)$, $Sp(k)$ and $U(n)$ of $Sp(n+k)$ in (6.1). If we take $S^r(\ )$ to be the rth symmetric product, then the complexified Lie algebra $\mathfrak{sp}(n+k)$ may be identified with $S^2(E+G)$, and $E = K + \bar{K}$.

Let $(T'_zZ)^h$ (resp. $(T'_zZ)^v$) denote the horizontal (resp. vertical) holomorphic tangent spaces at $z \in Z$. Then following [10], we deduce that

$$(6.2) \quad \begin{cases} (T'_zZ)^h \cong G \otimes K \\ \\ (T'_zZ)^v \cong S^2(K) \end{cases}$$

and $(TG_k(\mathbb{H}^{n+k}))^{\mathbb{C}} \cong G \otimes_{\mathbb{C}} E$.

The group $Sp(1) \cong SU(2)$ of unit quaternions acts naturally on $\mathbb{C}^2$, and, for $p = 2n + 2k-1$, it acts likewise on $S^p(\mathbb{C}^2) \cong$ {homogeneous polynomials of degree p in two complex variables} $\cong \mathbb{C}^{2n+2k} \cong \mathbb{H}^{n+k}$. We thus obtain an inclusion

(6.3) $\quad Sp(1) \longrightarrow Sp(n+k)$.

Consider the identification $\mathbb{C}P^1 \cong Sp(1)/U(1)$ and let $\eta \cong \mathbb{C}$ be the basic representation of the subgroup $U(1)$ of $Sp(1)$. We may define an inclusion $\Psi:U(1) \longrightarrow Sp(n+k)$, via the assignments

$$(6.4) \quad \begin{cases} E + G = S^p(\eta+\bar{\eta}) = \eta^p + \eta^{p-2} + \ldots + \bar{\eta}^{p-2} + \bar{\eta}^p \\ K \qquad = \eta^p + \eta^{p-2} + \ldots + \eta^{2k+1} \\ G \qquad = \eta^{2k-1} + \ldots + \eta + \bar{\eta} + \ldots + \bar{\eta}^{2k-1} \end{cases}$$

Relative to the standard inclusion $i:U(1) \longrightarrow Sp(1)$, the basic representation of $Sp(1)$ on $\mathbb{C}^2$ decomposes as $\eta \oplus \bar{\eta}$, and the inclusion (6.3) relative to Ψ, induces a map

(6.5) $\quad \psi : \mathbb{C}P^1 \longrightarrow Z$.

We also have the complexified Lie algebra decomposition $\mathfrak{sp}(1) = \mathfrak{u}(1) \oplus (\eta^2 \oplus \bar{\eta}^2)$, and on taking $\eta^2 \cong T'\mathbb{C}P^1$ at the identity coset, we find that $\psi_*(\eta^2) \subset G \otimes K$. This implies that ψ is holomorphic and π-horizontal with respect to (6.1). Hence $\phi = \pi\circ\psi$ defines an isotropic harmonic map $\phi:\mathbb{C}P^1 \to G_k(\mathbb{H}^{n+k})$.

Interchanging the roles of k and n above, and then setting $k = 1$, we recover the Veronese map $\psi:\mathbb{C}P^1 \to \mathbb{C}P^{2n+1}$.

EXAMPLE 6.2. Now let us recall the homogeneous fibration (4.2) over the real oriented Grassmannian

$$(6.6) \quad S^0_{k,r} \cong SO(2r+k)/U(r) \times SO(k) \longrightarrow G^0_k(\mathbb{R}^{2r+k}) \cong SO(2r+k)/SO(2r) \times SO(k) .$$

Again, to simplify the writing let us denote $S^0_{k,r}$ by Z.

Let $E \cong \mathbb{C}^r$, $H \cong \mathbb{C}^k$ and $K \cong \mathbb{C}^{2r}$ denote the respective basic representation of the (complexified) subgroups $U(r)$ and $SO(k)$ of $SO(2r+k)$ in (6.6) with $E \oplus \bar{E} = K$. Using the fact that $\mathfrak{so}(2r+k) \cong \Lambda^2(K \oplus H)$, $E \otimes \bar{E} \cong \mathfrak{u}(r)$ and $\Lambda^2 H \cong \mathfrak{so}(k)$, we deduce that (pointwise)

$$(6.7) \quad \begin{cases} (T'Z)^h \cong E \otimes H \\ (T'Z)^v \cong \Lambda^2 E . \end{cases}$$

Henceforth, we assume k to be even. The group $SO(3) \cong SU(2)$ also acts naturally on $\mathbb{C}^2$ and likewise on $S^q(\mathbb{C}^2) \cong \mathbb{C}^{2r+k}$ where $q = 2r+k-1$, and where we consider the representations to be real; hence we define an inclusion

(6.8) $\quad SO(3) \longrightarrow SO(2r+k)$.

The technique of Example 6.1. can then be applied as follows. We have

$$S^q(\eta \oplus \bar{\eta}) = \eta^q \oplus \eta^{q-2} \oplus \ldots \oplus \eta \oplus \bar{\eta} \oplus \ldots \oplus \bar{\eta}^q \cong E \oplus \bar{E} \oplus H .$$

In this case, we can define a map $\phi: S^2 \longrightarrow Z$ by an assignment for H (and therefore for E) in terms of η: define

$$H = \bigoplus_{i=0}^{k/2-1} (\eta^{\alpha_i} \oplus \bar{\eta}^{\alpha_i})$$

where for $0 \le i \le k/2 - 1$, $\alpha_i = 2i + 1$. Then ψ is holomorphic and π-horizontal and thus $\phi = \tilde{\pi} \circ \psi$ defines an isotropic harmonic map $\phi: S^2 \longrightarrow G_k^0(\mathbb{R}^{2r+k})$, for k even.

BIBLIOGRAPHY

1. R.L. Bryant, 'Lie groups and twistor spaces', preprint, Rice University (1983)

2. E. Calabi, 'Minimal immersions of surfaces in Euclidean spheres', J. Diff. Geom. 1, (1967), 111-125.

3. S.S. Chern, 'On minimal spheres in the four-sphere', Studies and essays presented to Y.W. Chen in the Math. Res. Centre, Taiwan Nat. Univ., Taipei (1970), 137-150.

4. J. Eells and L. Lemaire, 'A report on harmonic maps', Bull. London Math. Soc. 10 (1978), 1-68.

5. J. Eells and J.C. Wood, 'Harmonic maps from Riemann surfaces to complex projective spaces', Adv. in Math. Vol. 49, No. 3, (1983), 217-263.

6. S. Erdem and J.C. Wood, 'On the construction of harmonic maps into a Grassmannian', J. London Math. Soc. (2), 28 (1983), 161-174.

7. J.F. Glazebrook, 'Isotropic harmonic maps to Kähler manifolds and related properties', Ph.D. Thesis, University of Warwick (1983).

8. J.F. Glazebrook, 'The construction of a class of harmonic maps to quaternionic projective space', J. London Math. Soc. (2) 30(1984) 151-159.

9. I.R. Porteous, Topological Geometry, Cambridge University Press (1981).

10. S.M. Salamon, 'Harmonic and holomorphic maps', to appear in 'Seminar Luigi Bianchi', Springer Lecture Notes.

11. H.H. Wu, The equidistribution theory of holomorphic curves, Ann Math. Studies 64, Princeton University Press (1970).

DEPARTAMENTO DE MATEMATICAS
CINVESTAV - IPN
APARTADO POSTAL 14-740
07000 MEXICO 14, D.F.

Contemporary Mathematics
Volume 49, 1986

CHARACTERIZING $\mathbb{CP}_n$ BY THE SPECTRUM OF THE LAPLACIAN

S. I. Goldberg[1]

ABSTRACT. Complex projective space with the Fubini metric is characterized by the spectrum of the Laplacian on 2-forms in all dimensions.

1. INTRODUCTION. The implications of the spectrum of the Laplacian of a smooth compact Riemannian manifold on its geometry have been extensively studied during the past twenty years. The Euler number and signature of a manifold are determined by the spectrum. Patodi [13] showed that it is possible to determine whether the metric of a real Riemannian manifold is of constant curvature or Einstein from the spectra of its Laplacian on p-forms for all $p = 0,1,\ldots,\dim M$. For hermitian manifolds (M,g), the arithmetic genus is determined by its complex Laplacian Δ^c acting on forms of bidegree (q,r). Gilkey [6] proved that it is possible to determine whether g is a Kaehler metric from a knowledge of the spectra of Δ^c, and subsequently with Sacks [7] showed that if g is a Kaehler metric, then it is possible to determine whether it has constant holomorphic curvature from its spectra. Since the only compact manifold with a metric of constant positive holomorphic curvature is complex projective space $\mathbb{CP}_n$ with the Fubini-Study metric g_0 this proves that if (M,g) and $(\mathbb{CP}_n, g_0)$ are isospectral, then they are holomorphically isometric.

Now, let (M,J,g) be a compact connected Kaehler manifold with complex structure J and Kaehler metric g, and denote by $\Delta = -(dd^* + d^*d)$, where d

---

1980 Mathematics Subject Classification: 53C55, 58G25.

[1]Supported by Natural Sciences and Engineering Research Council of Canada.

is the operator of exterior differentiation and d^* is its adjoint with respect to the Riemannian metric g, the real Laplacian acting on p-forms. Then, for each $p = 0,1,2,\ldots$, we have the spectrum of Δ:

$$\mathrm{Spec}^p(M,g) = \{0 \geqq \lambda_{1,p} \geqq \lambda_{2,p} \geqq \cdots \geqq \lambda_{k,p} \geqq \cdots \downarrow -\infty\},$$

each eigenvalue being repeated as many times as its multiplicity. Hodge theory implies that $0 \in \mathrm{Spec}^p(M,g)$ if and only if the pth Betti number $b_p(M)$ is not zero, and its multiplicity is then $b_p(M)$.

If $\mathrm{Spec}^p(M,g) = \mathrm{Spec}^p(\mathbb{CP}_n,g_0)$ for a fixed value of p, is (M,J,g) holomorphically isometric with $(\mathbb{CP}_n,J_0,g_0)$, where J_0 is the standard complex structure of $\mathbb{CP}_n$? The answer is yes in the following cases:

1. $(M,J) = (\mathbb{CP}_n,J_0)$ and $p = 0$ [1], [11];
2. $p = 0$ and $n \leqq 6$ [14];
3. $p = 1$ and $8 \leqq n \leqq 51$ [15];
4. $p = 2$ and $n \neq 2, 8$ [5][1].

The last case is a vast improvement of the theorem of Gilkey and Sacks. In this paper, we show that the answer is yes in the exceptional dimensions of case 4 as well, thereby completely characterizing $(\mathbb{CP}_n,J_0,g_0)$ by $\mathrm{Spec}^2(M,g)$ for all n. This is the only case known where the geometry of (M,g) is completely determined by $\mathrm{Spec}^p(M,g)$ for some fixed p and in all dimensions, and yields an affirmative reply to the question asked in [5].

THEOREM. Let (M,J,g) be a Kaehler manifold with $\mathrm{Spec}^2(M,g) = \mathrm{Spec}^2(\mathbb{CP}_n, g_0)$. Then, (M,J,g) is holomorphically isometric to $(\mathbb{CP}_n,J_0,g_0)$ in all dimensions.

2. KAEHLER GEOMETRY. Let (M,J,g) be a Kaehler manifold. In terms of local complex coordinates, $g = 2\,\Sigma\, g_{ij^*}dz^i \otimes d\bar{z}^j$ and the Ricci tensor $S = 2\,\Sigma\, R_{ij^*}dz^i \otimes d\bar{z}^j$, $i,j,\ldots = 1,\ldots,n, i^* = i + n$, where the coefficients R_{ij^*} are given in terms of the components $R^i_{jk\ell^*}$ of the Riemann curvature

[1]The proof of Theorem 1 in [5] breaks down in dimension 2.

tensor by $-\Sigma R^k_{ikj*}$ (see [7]). Denote the scalar curvature by $\rho = 2\,\Sigma R_{ii*}$, and by $|R|$ and $|S|$ the norms of R and S, respectively, that is, $|R|^2 = 4\Sigma|R^i_{jk\ell*}|^2$ and $|S|^2 = 2\Sigma|R_{ij*}|^2$.

Define the scalars $\rho_1,\ldots,\rho_n$ by

$$\det(\delta_{ij} + tR_{ij*}) = \sum_{k=0}^{n} \binom{n}{k} \rho_k t^k .$$

Then, $\rho_0 = 1$, $\rho_1 = \rho/2n$, $\rho_2 = (\rho^2 - 2|S|^2)/4n(n-1),\ldots,\rho_n = \det(R_{ij*})$.

Let ω and c_1, respectively, be the fundamental and first Chern classes of M. Then,

LEMMA [12]. *If* M *is cohomologically Einstein, that is, if* $c_1 = a\omega$ *for some real number* a,

$$\int_M \rho_k * 1 = (2\pi a)^k \int_M * 1 ,$$

where $*1$ *is the volume element.*

PROOF. Since ω is the cohomology class of the Kaehler form $\Omega = \sqrt{-1}\,\Sigma g_{ij*} dz^i \wedge d\bar{z}^j$ and the Ricci 2-form $\gamma = (\sqrt{-1}/2\pi)\Sigma R_{ij*} dz^i \wedge d\bar{z}^j$ is a representative of c_1, $\gamma = a\Omega + d\eta$. Therefore,

$$\gamma^k = a^k \Omega^k + \sum_{\ell=1}^{k} C(\ell)\,\Omega^{k-\ell} \wedge (d\eta)^\ell ,$$

where $C(\ell)$ is a constant. Let Λ denote the interior product by Ω. Then, $\Lambda^k \Omega^k = k!n!/(n-k)!$ and $\Lambda^k \gamma^k = (k!)^2 \binom{n}{k} \rho_k/(2\pi)^k$. Hence,

$$\rho_k = (2\pi a)^k + \sum_{\ell=1}^{k} D(k,\ell)\,\Lambda^\ell (d\eta)^\ell ,$$

where $D(k,\ell)$ is a constant. Since

$$\int_M \Lambda^{\ell}(d\eta)^{\ell} * 1 = \int_M (d\eta)^{\ell} \wedge * \Omega^{\ell}$$
$$= \int_M d(\eta \wedge d\eta)^{\ell-1} \wedge * \Omega^{\ell}$$
$$= \int_M (\eta \wedge d\eta)^{\ell-1} \wedge * \delta\Omega^{\ell}$$
$$= 0 ,$$

the lemma follows.

3. PROOF OF THEOREM. For $p = 2$, the Minakshisundaram-Pleijel-Gaffney formula is given by

$$\sum_{k=0}^{\infty} \exp(\lambda_{k,2}\, t) = \frac{1}{(4\pi t)^n} \sum_{i=0}^{N} a_{i,2}\, t^i + 0(t^{N-n+1}), \quad t \downarrow 0 ,$$

where (cf. Patodi [13]) the coefficients $a_{i,2}$, $i = 0, 1, 2,$ are given by

(3.1) $$a_{0,2} = n(2n - 1) \int_M * 1 ,$$

(3.2) $$a_{1,2} = \frac{2n^2 - 13n + 12}{6} \int_M \rho * 1$$

and

(3.3) $$a_{2,2} = \frac{1}{360} \int_M \{5(2n^2 - 25n + 60)\rho^2 - 2(2n^2 - 181n + 540)|S|^2 + 2(2n^2 - 31n + 120)|R|^2\} * 1 .$$

Since $\mathrm{Spec}^2(M,g) = \mathrm{Spec}^2(\mathbb{CP}_n, g_0)$, we have

(3.4) $$\dim_{\mathbb{C}} M = n, \quad b_2(M) = b_2(\mathbb{CP}_n) = 1 ,$$

(3.5) $$\int_M * 1 = \int_{\mathbb{CP}_n} * 1, \quad \int_M \rho * 1 = \int_{\mathbb{CP}_n} \rho' * 1 ,$$

(3.6) $$a_{2,2} = a'_{2,2} ,$$

where the prime indicates the corresponding quantities in $(\mathbb{CP}_n', J_0, g_0)$.

The second part of (3.4) says that M is cohomologically Einstein. Thus, by the Lemma,

(3.7) $$\int_M \rho * 1 = 4n\pi a \int_M * 1 ,$$

(3.8) $$\int_M (\rho^2 - 2|S|^2) * 1 = 16n(n-1)\pi^2 a^2 \int_M * 1 ,$$

from which

(3.9) $$(n-1)(\int_M \rho * 1)^2 = n(\int_M * 1) \int_M (\rho^2 - 2|S|^2) * 1 .$$

Since $(\mathbb{CP}_n, J_0, g_0)$ has constant holomorphic curvature, say H,

(3.10) $$\rho' = n(n+1)H, \quad |S'|^2 = n(n+1)^2 H^2/2, \quad |R'|^2 = 2n(n+1)H^2 .$$

From (3.5), (3.7) and (3.10), we find that $a = (n+1)H/4\pi$. Substituting from (3.10) into (3.3), then for H in terms of a, and finally for a from (3.7), we get

$$a'_{2,2} = \frac{1}{360n(n+1)} (10n^4 - 117n^3 + 362n^2 - 183n - 60) \frac{(\int_M \rho * 1)^2}{\int_M * 1} .$$

Formula (3.9) then yields

$$360(n^2-1)a'_{2,2} = (10n^4 - 117n^3 + 362n^2 - 183n - 60) \int_M (\rho^2 - 2|S|^2) * 1 ,$$

so, from (3.6)

$$\begin{aligned} \int_M \{(n^2-1)(2n^2 - 31n + 120)|R|^2 \\ + 4(2n^4 + 16n^3 - 44n^2 - 91n + 120)|S|^2 \\ -2(2n^3 + 18n^2 - 77n + 60)\rho^2\} * 1 = 0 \end{aligned}$$

which may be written in the form

(3.11) $$\begin{aligned} (n^2-1)(n-8)(2n-15) \int_M \{|R|^2 - \frac{2}{n(n+1)} \rho^2\} * 1 \\ + 4(2n^4 + 16n^3 - 44n^2 - 91n + 120) \int_M (|S|^2 - \frac{\rho^2}{2n}) * 1 = 0 . \end{aligned}$$

Consider the projective curvature tensor W constructed in [8] (see also [7]):

$$W^i_{jk\ell^*} = R^i_{jk\ell^*} + \frac{1}{n+1}(R_{j\ell^*}\delta^i_k + R_{k\ell^*}\delta^i_j) .$$

This tensor vanishes if and only if the Kaehler metric has constant holomorphic curvature. Setting $|W|^2 = 4\Sigma|W^i_{jk\ell^*}|^2$, we obtain

$$0 \leqq |W|^2 = |R|^2 - \frac{4}{n+1}|S|^2 ,$$

so since $|S|^2 \geqq \rho^2/2n$, it follows that $|R|^2 \geqq 2\rho^2/n(n+1)$, and (3.11) then yields $|S|^2 = \rho^2/2n$, $|R|^2 = 2\rho^2/n(n+1)$ for $n \neq 2, 8$. Thus, (M,J,g) has constant holomorphic curvature H_1. From (3.5), $H_1 = H$, so, for $n \neq 2, 8$, (M,J,g) is holomorphically isometric with $(\mathbb{CP}_n, J_0, g_0)$.

<u>The case</u> $n = 2$: Since $b_2(M) = 1$, the dimension of the space $H^{1,1}(M,R)$ of harmonic forms of bidegree $(1,1)$ is 1. Multiplying the Kaehler metric g by some constant, if necessary, we may therefore assume that g is a Hodge metric. Thus, ω is a positive element of $H^{1,1}(M,Z)$, and since $c_1 = ((n+1)H/4\pi)\omega$, the first Chern class is positive. Consequently, by a Theorem of Kobayashi [9], $b_1(M) = 0$ (see also [8]). It follows that the Euler characteristic $\chi(M) = \chi(\mathbb{P}\,\mathbb{C}_2)$. From the Gauss-Bonnet theorem

$$\chi(M) = \frac{1}{32\pi^2}\int_M (|R|^2 - 4|S|^2 + \rho^2) * 1 .$$

Thus,

$$\int_M (|R|^2 - 4|S|^2 + \rho^2) * 1 = \int_{\mathbb{CP}_2} (|R'|^2 - 4|S'|^2 + \rho'^2) * 1 . \tag{3.12}$$

Formula (3.9) yields by virtue of (3.1) and (3.2)

$$\int_M (2|S|^2 - \rho^2) * 1 = \int_{\mathbb{CP}_2} (2|S'|^2 - \rho'^2) * 1 . \tag{3.13}$$

Thus,

$$\int_M (|R|^2 - |R'|^2) * 1 = \int_M (\rho^2 - \rho'^2) * 1 . \tag{3.14}$$

From (3.3)

$$360a_{2,2} = \int_M (90\rho^2 - 372|S|^2 + 132|R|^2) * 1 ,$$

so since $a_{2,2} = a'_{2,2}$,

$$(3.15) \quad 132 \int_M (|R|^2 - |R'|^2) * 1 + 90 \int_M (\rho^2 - \rho'^2) * 1$$
$$= 372 \int_M (|S|^2 - |S'|^2) * 1 .$$

Substituting for $|S|^2 - |S'|^2$ from (3.13) in (3.15), we obtain

$$132 \int_M (|R|^2 - |R'|^2) * 1 = 96 \int_M (\rho^2 - \rho'^2) * 1 .$$

Applying (3.14),

$$\int_M |R|^2 * 1 = \int_{\mathbb{C}\mathbb{P}_2} |R'|^2 * 1 , \quad \int_M \rho^2 * 1 = \int_{\mathbb{C}\mathbb{P}_2} \rho'^2 * 1$$

and so, from (3.13),

$$\int_M |S|^2 * 1 = \int_{\mathbb{C}\mathbb{P}_2} |S'|^2 * 1 .$$

Thus,

$$\int_M |W|^2 * 1 = \int_M (|R|^2 - \frac{4}{3} |S|^2) * 1$$
$$= \int_{\mathbb{C}\mathbb{P}_2} (|R'|^2 - \frac{4}{3} |S'|^2) * 1$$
$$= \int_{\mathbb{C}\mathbb{P}_2} |W'|^2 * 1$$
$$= 0 .$$

We conclude that $W = 0$, that is, (M,J,g) is holomorphically isometric with $(\mathbb{C}\mathbb{P}_2, J_0, g_0)$.

This can also be seen more easily and directly by substituting for $|R|^2$ in (3.11) from formula (4.4) in [14] as was pointed out to us by Vanhecke after this paper was completed.

The case $n = 8$: From (3.11),

$$\int_M \left(|S|^2 - \frac{\rho^2}{2n}\right) * 1 = 0$$

for $n = 8$, so since $|S|^2 \geqq \rho^2/2n$, equality holds, and consequently g is an Einstein metric with positive scalar curvature by (3.5). By Theorem 5 of Kobayashi [10], a principal circle bundle P over M with projection π may be constructed having an Einstein metric with positive scalar curvature $k = (2n+1)\rho/(2n+2)$. (The construction of this bundle is natural in the sense that P is a space of constant positive curvature if and only if M is a space of constant positive holomorphic curvature.)

From Cor. 4 on p. 256 in [2] the volume of P is bounded above by the volume of the sphere $S^{2n+1}(k/2n)$ having constant curvature $k/2n$, that is,

$$v(P) \leqq v(S^{2n+1}(k/2n)) , \tag{3.16}$$

with equality attained only if P has constant curvature, that is, only if P is isometric with $S^{2n+1}(k/2n)$.

We now relate $v(M)$ and $v(P)$ (see [3]), the metric of P being given by $G = \pi * g + b^2\beta^2$, where β is the 1-form on P defining a connection in P and b is a positive number. Let X_0 be the vector field induced on P by the action of S^1, so $\beta(X_0) = 1$ and X_0 is vertical. The integral curves of X_0 are periodic with period 2π. Thus, the length of the fibres of P with respect to the metric G is $2\pi b$. Since P is approximated locally by the Riemannian product $S^1 \times U$, where U is a small open set of M and S^1 is a circle of radius b,

$$2\pi b\ v(M) = v(P) . \tag{3.17}$$

Applying this to the case $P = S^{2n+1}(k/2n)$ and $M = \mathbb{CP}_n(H)$ which has constant holomorphic curvature $H = \rho/n(n+1)$, we obtain

$$2\pi b\ v(\mathbb{CP}_n(H)) = v(S^{2n+1}(k/2n)) . \tag{3.18}$$

Substituting (3.17) and (3.18) in the inequality (3.16), the factors $2\pi b$ cancel, and we get

$$v(M) \leqq v(\mathbb{CP}_n(H)) .$$

Equality is attained only if P is isometric with $S^{2n+1}(k/2n)$, and hence only if (M,g) is isometric with $\mathbb{CP}_n(H)$, so by (3.5), (M,J,g) is holomorphically isometric with $(\mathbb{CP}_n, J_0, g_0)$.

The case $n = 1$ is established in the same manner, and is also an easy consequence of Schwarz's inequality together with (3.3), (3.5) and (3.6).

REMARKS. (a) It was recently shown [4] that if $\mathrm{Spec}^{1,1}(M,g) = \mathrm{Spec}^{1,1}(\mathbb{CP}_n, g_0)$, $n \neq 2$, then (M,J,g) is holomorphically isometric with $(\mathbb{CP}_n, J_0, g_0)$. Let $h^{q,r}$ denote the dimension of the space of harmonic forms of bidegree (q,r). Then, if $\mathrm{Spec}^{1,1}(M,g) = \mathrm{Spec}^{1,1}(\mathbb{CP}_2, g_0)$, $h^{1,1} = 1$. Thus, $c_1 = a\omega$, $a > 0$, from which $h^{1,0} = h^{0,1} = 0$. It follows that $\chi(M) = \chi(\mathbb{CP}_2)$. Therefore, as before, (M,J,g) is holomorphically isometric with $(\mathbb{CP}_2, J_0, g_0)$, and so Theorem 1 in [4] is true in all dimensions.

(b) Suppose $\mathrm{Spec}^p(M,g) = \mathrm{Spec}^p(M_0, g_0)$, $p = 0,1,2$, where (M_0, g_0) is a spherical space form. Is M isometric to M_0? The known cases are $M_0 = S^n$ and $\mathbb{RP}_n$. The next case of interest is that of a generalized lens space $L(k;q_1,\ldots,q_n) = G_0 \backslash S^{2n-1}$, where $S^{2n-1} = \{(z_1,\ldots,z_n | \Sigma |z_i|^2 = 1\}$ and $G_0 \approx Z_k$ acts by $(z_1,\ldots,z_n) \to (\zeta^{q_1} z_1, \ldots, \zeta^{q_n} z_n)$, $\zeta = e^{2\pi i/k}$, $(q_j, k) = 1$, $j = 1,\ldots,n$. If $M_0 = L(k;1,1,\ldots,1)$, then M is isometric to M_0. This fact is due to R. Kulkarni. However, it is by no means clear that $\mathrm{Spec}^o(M)$ would alone characterize M. Are there Milnor-type examples for 3- or 5-dimensional lens spaces? For more rigid spaces such as the Poincaré homology sphere less information may suffice.

(c) The Hopf fibration $S^1 \to S^{2n+1} \to \mathbb{CP}_n$ may be useful as the author's paper with T. Ishihara [J. Differential Geometry 13 (1978), 139-144] on Riemannian submersions commuting with the Laplacian indicates. Indeed, spectral information of $(\mathbb{CP}_n, g_0)$ can be pulled back to S^{2n+1}.

BIBLIOGRAPHY

1. M. Berger, P. Gauduchon and E. Mazet, Le spectre d'une variété riemanniennes, Lecture Notes in Math., vol. 194, Springer-Verlag, Berlin and New York, 1971.

2. R. L. Bishop and R. J. Crittenden, "Geometry of manifolds," Academic Press, New York and London, 1964.

3. R. L. Bishop and S. I. Goldberg, On the topology of positively curved Kaehler manifolds II, Tôhoku Math. J. 17 (1965), 310-318.

4. B.-Y. Chen and K. Ogiue, On the spectrum of $\square^{1,1}$ on Kaehler manifolds, Arch. Math. 40 (1983), 367-371.

5. B.-Y. Chen and L. Vanhecke, The spectrum of the Laplacian of Kaehler manifolds, Proc. Amer. Math. Soc. 78 (1980), 82-86.

6. P. Gilkey, The spectral geometry of real and complex manifolds, Proc. Symposium Pure Math., Amer. Math. Soc. 27, Part 2 (1975), 265-280.

7. P. Gilkey and J. Sacks, Spectral geometry and manifolds of constant holomorphic sectional curvature, ibid. 27, Part 2 (1975), 281-285.

8. S. I. Goldberg, "Curvature and homology," Academic Press, New York and London, 1962.

9. S. Kobayashi, On compact Kaehler manifolds with positive definite Ricci tensor, Ann. of Math. 74 (1961), 570-574.

10. __________, Topology of positively pinched Kaehler manifolds, Tôhoku Math. J. 15 (1963), 121-139.

11. A. Lascoux and M. Berger, Variétés Kählériennes compactes, Lecture Notes in Math., vol. 154, Springer-Verlag, Berlin and New York, 1970.

12. K. Ogiue, Generalized scalar curvatures of cohomological Einstein Kaehler manifolds, J. Differential Geometry 10 (1975), 201-205.

13. V. K. Patodi, Curvature and the fundamental solution of the heat operator, J. Indian Math. Soc. 34 (1970), 269-285.

14. S. Tanno, Eigenvalues of the Laplacian of Riemannian manifolds, Tôhoku Math. J. 25 (1973), 391-403.

15. __________, The spectrum of the Laplacian for 1-forms, Proc. Amer. Math. Soc. 45 (1974), 125-129.

Department of Mathematics
University of Illinois
Urbana, Illinois 61801

Contemporary Mathematics
Volume 49, 1986

STABLE MINIMAL SURFACES IN FLAT TORI

Mario J. Micallef[1]

1. INTRODUCTION AND STATEMENT OF MAIN RESULT

A periodic surface in $\mathbb{R}^n$ is a surface which is invariant under the group of translations generated by a lattice Λ in $\mathbb{R}^n$. The quotient surface lies in the flat torus $T^n = \mathbb{R}^n/\Lambda$. The study of periodic minimal surfaces in $\mathbb{R}^3$ was started by H. A. Schwarz in 1890 [S]. More recently, Meeks [Me], Nagano and Smyth [N-S1], [N-S2] and others have made more extensive study of such minimal surfaces in $\mathbb{R}^n$. In this note we study the stability of minimal surfaces in flat tori and prove the following

THEOREM: Let $F\colon M^2 \to T^n$ be a stable minimal immersion of a closed (i.e. compact and without boundary) oriented surface M^2. Suppose that the Gauss map G of F factors through the two-sphere, i.e. we have the commutative diagram

$$\begin{array}{ccc} M^2 & \xrightarrow{\;G\;} & Q_{n-2} \subset \mathbb{C}P^{n-1} \\ & {\scriptstyle h}\searrow \quad \nearrow{\scriptstyle \ell} & \\ & \mathbb{C}P^1 & \end{array}$$

for some holomorphic maps h and ℓ (G is holomorphic because F is minimal where holomorphic is meant with respect to the complex structure defined by means of isothermal parameters for the metric on M induced by F and the orientation on M^2. Then the image of F lies in an even dimensional totally geodesic subtorus T^{2m} of T^n and F is holomorphic with respect to some orthogonal complex structure on T^{2m}. Furthermore, if $m < g$ where g is the genus of M^2, then the Jacobian variety of M^2 is not simple.

1980 A.M.S. Subject Classification 53A10, 58E12, 53C42, 49F10
[1] Research partially supported by N.S.F. grant DMS 8401930.

REMARKS: (1) For stable minimal immersions $F: M^2 \to T^4$, the above theorem is true without any assumption on the Gauss map (Theorem 1' in [Mi]). The proof then, however, is quite different.

(2) It is not known whether the above theorem holds for periodic minimal surfaces in $\mathbb{R}^n (n \geq 4)$, for the stability of the periodic surface is not known to imply the stability of the corresponding compact surface in T^n (and vice versa).

(3) Both the above theorem and Theorem 1' in [Mi] have content only for surfaces of genus greater than or equal to three. For genus zero and one, this is obvious. It follows easily from the Gauss-Bonnet theorem (see [Me] and [N-S2]), that it is impossible to minimally immerse a surface of genus two into T^3. Since a surface of genus g has g linearly independent holomorphic differentials, such a surface can be fully and minimally immersed in a flat torus of real dimension at most 2g. It is then easy to see that a surface of genus two immerses fully and minimally only into T^4, and then only by the Albanese mapping (which is holomorphic) into its Jacobian variety. It is also worth pointing out that a surface of genus three which lies fully and minimally in T^6 is automatically holomorphic if it is not hyperelliptic. (This result was also independently arrived at by William H. Meeks, III.)

Finally, it is a pleasure to thank William H. Meeks, III for many stimulating discussions.

2. STABLE HYPERELLIPTIC MINIMAL SURFACES

Recall that a Riemann surface is hyperelliptic if and only if it can be realized as a two-sheeted branched covering of $\mathbb{C}P^1$.

COROLLARY 1: Let $F: M^2 \to T^n$ be a stable minimal immersion of a closed oriented surface M^2. Suppose that M^2 is hyperelliptic in the conformal structure induced by the first fundamental form of F and the orientation on M^2. Then the conclusions of the theorem in §1 hold.

PROOF: The Gauss map G of F is that of the corresponding periodic minimal surface in $\mathbb{R}^n$ (whose immersion we still denote by F). Let z be a local complex co-ordinate on M^2. Then $(F^1_z, \ldots, F^n_z)$ are homogeneous co-ordinates for G(z) and F minimal implies that $\partial F^1 = F^1_z dz, \ldots, \partial F^n = F^n_z dz$ are n holomorphic differentials on M^2. Let $i: M^2 \to \mathbb{C}P^{g-1}$ be the canonical mapping of the Riemann surface

M^2 whose genus is g. Then there exists a complex linear map $\lambda: \mathbb{C}P^{g-1} \to \mathbb{C}P^{n-1}$ such that $G = \lambda \circ i$. Now i factors through $\mathbb{C}P^1$ if and only if M^2 is hyperelliptic and therefore G factors through $\mathbb{C}P^1$ if M^2 is hyperelliptic. The corollary then follows immediately from the theorem in §1.

The corollaries 2 and 2' below follow immediately from Corollary 1 and remark 1 in §1.

COROLLARY 2: The only conformal stable minimal immersion into a flat torus of a hyperelliptic Riemann surface whose Jacobian variety is simple is the Albanese mapping of the surface into its Jacobian variety.

COROLLARY 2': A Riemann surface of genus greater than or equal to three whose Jacobian variety is simple cannot be conformally immersed as a stable minimal surface in the flat four-real dimensional torus.

3. PROOF OF THEOREM

We first establish our notation. Since the tangent bundle of T^n is geometrically trivial, we may regard sections of the complexified tangent and normal bundles of M (denoted by $T_{\mathbb{C}}M$ and $N_{\mathbb{C}}M$ respectively) as $\mathbb{C}^n$ valued functions. Covariant differentiation by means of the usual Levi Civita (flat) connection on the tangent bundle of T^n will be denoted by d when it acts on $\mathbb{C}^n$ valued functions. We will also use $\partial_z (\partial_{\bar{z}})$ instead of $d_z (d_{\bar{z}})$ when we differentiate $\mathbb{C}^n$ valued functions in the direction $\partial/\partial z$ $(\partial/\partial \bar{z})$ along the oriented surface M^2 immersed in T^n, where z is a local complex co-ordinate on the surface.

It is shown in §2 of [Mi], that stability of M^2 is equivalent to the inequality

$$(1) \qquad \int_{M^2} |(\partial s)^T|^2 \le \int_{M^2} |(\bar{\partial} s)^N|^2$$

for all sections s of $N_{\mathbb{C}}M$. In (1), $\partial s = (\partial s^1, \ldots, \partial s^n)$ where $s = (s^1, \ldots, s^n)$ (similarly for $\bar{\partial} s$), the superscripts T and N denote orthogonal projection onto $T_{\mathbb{C}}M$ and $N_{\mathbb{C}}M$ respectively, $|(\partial s)^T|$ is the length of the $\mathbb{C}^n$ valued one form $(\partial s)^T$ with respect to the induced metric on M(similarly for $(\bar{\partial} s)^N$) and the integrations are carried out with respect to the element of area for the induced metric on M.

We now produce holomorphic sections of $N_{\mathbb{C}}M$ to plug into (1). (Recall that by a proposition of Koszul and Malgrange [K-M], a complex vector bundle with connection D over a Riemann surface can be made into a holomorphic vector bundle by defining the $\bar{\partial}$ operator to be the (0,1) part of D, that is, a section s is holomorphic if and only if $D_{\bar{z}}s=0$). We first note that over $Q_{n-2} = G_{2,n}$ we have the tautological 2-plane bundle $\gamma_{2,n}$. Let $\gamma_{2,n}^{\perp}$ be the orthogonal complement of $\gamma_{2,n}$ in $G_{2,n} \times \mathbb{R}^n$. The tangent and normal bundles of M^2 are geometrically isomorphic to $G^*(\gamma_{2,n})$ and $G^*(\gamma_{2,n}^{\perp})$ respectively. Since G factors through the two-sphere, $N_{\mathbb{C}}M = h^*(\eta_{\mathbb{C}})$ where h is as in the statement of the theorem, $\eta = \ell^*(\gamma_{2,n}^{\perp})$ and $\eta_{\mathbb{C}} = \eta \otimes_{\mathbb{R}} \mathbb{C}$. Viewing $\eta_{\mathbb{C}}$ as a holomorphic vector bundle (by the proposition of Koszul and Malgrange), a theorem of Grothendieck [G], then tells us that $\eta_{\mathbb{C}}$ splits as the direct sum of holomorphic line bundles $L_1 \oplus \dots \oplus L_{n-2}$. Let $L_1,\dots,L_p$ be the positive line bundles, $L_{p+1},\dots,L_r$ be the topologically trivial ones and $L_{r+1},\dots,L_{n-2}$ be all negative. By the Riemann-Roch theorem, we know that each of $L_1,\dots,L_p$, $L_{p+1},\dots,L_r$ admits a holomorphic section σ_j, $j \in \{1,\dots,r\}$. Let $s_j = \sigma_j \circ h$. Then s_j is a holomorphic section of $N_{\mathbb{C}}M$.

The argument now proceeds as in Theorem IV of [Mi]. We simply sketch it here for completeness and refer to [Mi] for details. By plugging s_j into (1) we first note that $(\partial s_j)^T=0$. Let $\langle\cdot,\cdot\rangle$ denote the usual Hermitian inner product on $\mathbb{C}^n$ which is conjugate linear in the second argument. Since $0 = C_1(\eta_{\mathbb{C}}) = C_1(L_1) + \dots + C_1(L_{n-2})$ only the following two cases arise:

<u>CASE (i)</u>: $L_1,\dots,L_{n-2}$ are <u>all trivial</u>. But then $(\partial s_j)^T = 0$ for all $j \in \{1,\dots,n-2\}$ and since $s_1,\dots,s_{n-2}$ span $N_{\mathbb{C}}M$ we have, for any section s of $N_{\mathbb{C}}M$, $(ds)^T = (\partial s)^T + \overline{(\partial \bar{s})^T} = 0$ where the bar denotes complex conjugation. Thus M is a totally geodesic two real-dimensional subtorus of T^n and, in particular, the theorem holds.

<u>CASE (ii)</u>: $1<r<n-2$. From now on we adopt the following range of indices:

$$1\le \mu,\nu\le p \qquad r+1 \le a \le n-2$$
$$1\le j,k\le r \qquad p+1 \le A \le n-2$$

Note that $\langle s_\mu,\bar{s}_j\rangle\equiv 0$ since $\langle s_\mu,\bar{s}_j\rangle$ is a holomorphic (and therefore constant) function on M and s_μ vanishes somewhere. This implies that $p\le n-2-r$. By considering meromorphic sections of $L_{r+1},\dots,L_{n-2}$ which have no zeroes we conclude that $p\ge n-2-r$. Therefore $p=n-2-r$

and the space orthogonal to the span of $\{\bar{L}_1,\ldots,\bar{L}_r\}$ is spanned by $L_1,\ldots,L_p$.

Using the fact that $(\partial s_j)^T=0$, it is easy to check that $\langle\partial s_j,\bar{s}_k\rangle$ is a holomorphic differential on M. Therefore $\langle\partial\sigma_j,\bar{\sigma}_k\rangle$ is a holomorphic differential on $\mathbb{C}P'$ and therefore identically zero. Thus $\partial:\Gamma(h^*(L_1)\oplus\ldots\oplus h^*(L_r))\to\Gamma((h^*(L_1)+\ldots+h^*(L_p))\otimes T^{1,0}M)$.

Let $\xi=h^*(L_1)\oplus\ldots\oplus h^*(L_r)\oplus T^{1,0}M$. The above considerations and the fact that F is minimal then imply that $d:\ \Gamma(\xi)\to\Gamma(\xi\otimes T^*M)$. Thus there exists a subspace Σ of complex dimension r+1 in $\mathbb{C}^n$ such that $\xi = M \times \Sigma$. Let $T = \Sigma \cap \bar{\Sigma}$ and t = complex dimension of T. Then $t = r-p$.

We now recall the following theorem from [Mi].

<u>THEOREM A</u>: Let $F:\ M^{2m}\to\mathbb{R}^{2n}$ be an immersion, with $\mathbb{R}^{2n}$ having the usual Euclidean metric. Suppose that there exist vector bundles E and V over M which satisfy conditions (i), (ii) and (iii) below:

(i) $T_{\mathbb{C}}M \cong E\oplus\bar{E},\quad N_{\mathbb{C}}M \cong V\oplus\bar{V}$

(ii) $E\oplus V$ is orthogonal to $\bar{E}\oplus\bar{V}$

(iii) $d:\ \Gamma(E\oplus V)\to\Gamma((E\oplus V)\otimes T^*M)$

Then there exist complex structures $\tilde{J}$ and J on M^{2m} and $\mathbb{R}^{2n}$ respectively such that $\tilde{J}$ is orthogonal with respect to the metric induced on M by the immersion, J is orthogonal with respect to the Euclidean inner product on $\mathbb{R}^{2n}$ and F is holomorphic with respect to $\tilde{J}$ and J.

We can now finish the proof of the theorem in this paper. If $t=0$, $r=p$ and the theorem follows from an application of Theorem A with $E = T^{1,0}M$ and $V = h^*(L_1)\oplus\ldots\oplus h^*(L_p)$. If $t\neq 0$, then since T is preserved by complex conjugation, $T = W\otimes_{\mathbb{R}}\mathbb{C}$ where W is a subspace of real dimension t in $\mathbb{R}^n$. It is then not hard to see that F(M) lies fully in a totally geodesic submanifold S of T^n which has real dimension $n-t = 2p+2$ (the lifts of S to $\mathbb{R}^n$ form a collection of affine subspaces orthogonal to W). S is isometric to $\mathbb{R}^b\times T^{n-t-b}$ with the product flat metric where $0\leq b\leq n-t$. Composing F with the projection onto the $\mathbb{R}^b$ factor gives b harmonic functions on M all of which must therefore be constant. But F(M) lies fully in S and therefore b=0, that is, S is a totally geodesic subtorus of T^n. It is also immediate that the normal bundle of M when M is viewed as a surface in S is equal to

$(M\times T)^{\perp}$ and that $(M\times T)^{\perp} = h^*(L_1)\oplus\ldots\oplus h^*(L_p)\oplus\overline{h^*(L_1)\oplus\ldots\oplus h^*(L_p)}$. The theorem in §1 is now proved by applying Theorem A with $E = T^{1,0}M$ and $V = h^*(L_1)\oplus\ldots\oplus h^*(L_p)$. (Note that $S=T^{2m}$ in the statement of the theorem and $m=p+1$).

Finally let $T^m_{\mathbb{C}}$ be the complex torus obtained from $S(=T^{2m})$ and the complex structure J with respect to which F is holomorphic. Then there exists holomorphic one-forms $\omega_1,\ldots,\omega_m$ which are linearly independent (since F is full) and such that $F\colon M^2 \to T^m_{\mathbb{C}}$ is given by $F(z) = \int_{z_0}^{z} (\omega_1,\ldots,\omega_m)$, z_0 some fixed point in M^2. If $m<g$, then we complete $\{\omega_1,\ldots,\omega_m\}$ to a basis and define $\tilde{F} = \int(\omega_1,\ldots,\omega_g)$. Then $\tilde{F}$ is the Albanese mapping and we see that $T^m_{\mathbb{C}}$ is a complex subtorus of the Jacobian variety of M^2, that is, the Jacobian variety of M^2 is not simple.

BIBLIOGRAPHY

[G] A. Grothendieck, Sur la classification des fibrées holomorphes sur la sphère de Riemann, Amer. J. Math. 79 (1957) 121-138.

[K-M] J. L. Koszul & B. Malgrange, Sur certaines fibrées complexes, Arch. Math. 9 (1958) 102-109.

[Me] W. H. Meeks, III, The conformal structure and geometry of triply periodic minimal surfaces in $\mathbb{R}^3$, Ph.D. Thesis, University of California, Berkeley, 1975.

[Mi] M. J. Micallef, Stable Minimal Surfaces in Euclidean Space, J. Diff. Geom. 19 (1984) 57-84.

[N-S1] T. Nagano & B. Smyth, Minimal Varieties and harmonic maps in Tori, Comm. Math. Helv. 50 (1975) 249-265.

[N-S2] T. Nagano & B. Smyth, Periodic minimal surfaces, Comm. Math. Helv. 53 (1978) 29-55.

[S] H. A. Schwarz, Bestimmung einer speziellen Minimalflächen usw., Preisschrift, in Gesammelte Mathematische Abhandlungen, Band I, Verlag von Julius Springer, Berlin, 1890, 6-91.

Department of Mathematics
3220 Angell Hall
University of Michigan
Ann Arbor, MI 48109-1003

Contemporary Mathematics
Volume 49, 1986

FOLIATION TECHNIQUES AND VANISHING THEOREMS

Ngaiming Mok[1]

ABSTRACT. Classically, vanishing theorems in Kähler geometry are proved using integral formulas or some form of the maximum principle. This usually involves some form of positivity or negativity of the curvature tensor. We examine in this article techniques involved in the situation of semipositive/seminegative curvature. It is our theory that the partial vanishing of tensors (or similar objects) under study can often be "integrated" to yield global proofs of vanishing theorems. For this it is necessary to study certain holomorphic foliations or families of complex submanifolds naturally associated with the problem.

INTRODUCTION. This is a general survey of global techniques used in the proofs of a variety of vanishing theorems on compact Kähler manifolds. Classically, the vanishing of certain cohomology groups of a Hermitian holomorphic vector bundle on a compact Kähler manifold is obtained by proving the vanishing of appropriate harmonic forms. This is done typically by a Bochner-Kodaira formula, an integral formula involving the curvature of the Hermitian vector bundle, the curvature of the underlying Kähler manifold and the gradient of the harmonic form. These formulas are obtained by directly computing $\Delta\|\omega\|^2$ for the harmonic form ω or by applying integration by parts to such pointwise computations. Such formulas were systematically investigated in Siu [20]. Our concern in the present article is to consider the situation when an application of such and similar integral formulas or some form of the

1980 Mathematics Subject Classification.
[1]Supported by Sloan Foundation.

maximum principle does not yield directly the vanishing of the tensors (e.g. harmonic forms) or similar objects under study. Namely, under certain conditions of semipositive or seminegative curvature, one may be able to obtain pointwise partial information on these objects. The global problem is then to "integrate" such partial information in order to prove the vanishing theorem desired. While there is no general method to deal with such situations a useful approach is to study holomorphic foliations or families of complex-analytic submanifolds specifically and naturally associated with the problem. In the present article we will discuss four different examples related to irreducible compact quotients of polydiscs and compact Hermitian symmetric manifolds.

The first example is an old theorem of Matsushima-Shimura [9] on the vanishing of certain cohomology groups attached to some seminegative Hermitian vector bundles on irreducible compact quotients of polydiscs $X = \Delta^n/\Gamma$, $n \geq 2$. Here X is said to be irreducible if no finite covering of X splits into a non-trivial Cartesian product of quotients of polydiscs. Since the irreducibility of X is needed for such vanishing theorems it is not possible to prove them directly from integral formulas. Nonetheless one can conclude from a standard Bochner-Kodaira formula the vanishing of certain components of harmonic forms φ under study. For any integer q satisfying $1 \leq q \leq n$ the Cartesian decomposition $\Delta^n = \Delta^q \times \Delta^{n-q}$ gives rise to a foliation $\tilde{\mathfrak{E}}$ of Δ^n by $\{z'\} \times \Delta^{n-2}$, $z' \in \Delta^q$. Assuming $\Gamma \subset (\mathrm{Aut}(\Delta))^n$, $\tilde{\mathfrak{E}}$ induces a foliation $\mathfrak{E}$ on X. The irreducibility of X implies that every leaf of $\mathfrak{E}$ is dense in X. The proof that φ vanishes on X identically can be obtained by studying the behavior of remaining components of φ along leaves of $\mathfrak{E}$.

The vanishing theorem of Matsushima-Shimura [9] yields in particular $H^1(X,T) = 0$ for the holomorphic tangent bundle T on an irreducible compact quotient X of Δ^n, $n \geq 2$, implying the local rigidity of such complex manifolds. Our second example is the strong rigidity of X, namely, the assertion that every compact Kähler manifold M diffeomorphic to X must be biholomorphic to X up to a change of orientation on X (i.e. taking conjugates in certain variables z_k on $\Delta^n(x_1,\ldots,z_n)$). This was proved independently by Jost-Yau [6],[7] in case of surfaces and by Mok [10] for all dimensions $n \geq 2$. Both works were based on the $\partial\bar{\partial}$-Bochner-Kodaira formula of Siu [18],[9], who proved the strong rigidity theorem for compact quotients of irreducible bounded symmetric domains of complex dimension ≥ 2. Let $f\colon M \to X = \Delta^n/\Gamma$ be a harmonic map homotopic to some diffeomorphism $f_0\colon M \to X$. Applying the $\partial\bar{\partial}$-Bochner-Kodaira formula of Siu it was proved (Jost-Yau [6]) that for the lifting $F\colon \tilde{M} \to \Delta^n$ to the universal covering manifolds the generic level sets of components F_k of F define on some open dense subset of $\tilde{M}$ an analytic foliation $\tilde{\mathcal{F}}_k$ by complex hypersurfaces. In [10] we study the induced foliation $\mathcal{F}_k$ on M. While a priori $\mathcal{F}_k$ is only defined outside a real-analytic subvariety of M which may disconnect the complement into several components, we showed using techniques of extension of complex analytic objects that $\mathcal{F}_k$ can be analytically continued to $M - V_k$ for some complex analytic subvariety V_k of codimension ≥ 2 in M. Still denoting the extended foliation by $\mathcal{F}_k$ we have generically $\mathcal{F}_k = f^*(\mathcal{E}_k)$ for some natural foliation $\mathcal{E}_k$ of X by projections of hyperdiscs. (The leaves of $_k$ are projections of $\Delta^{k-1} \times \{z_k\} \times \Delta^{n-k}$ into X via $\pi\colon \Delta^n \to X = \Delta^n/\Gamma$.) With some oversimplification the proof in Mok [10] was obtained by comparing the dynamical behavior of the two analytic foliations $\mathcal{F}_k$ and $\mathcal{E}_k$ on $M - V_k$ and X respectively. Namely, if F_k were neither holomorphic nor anti-holomorphic by studying level sets of $\lambda_k = |\partial F_k/\partial F_k|$ ($\partial\bar{F}_k$ and ∂F_k are proportional on $M - V_k$) we would be able to find a closed leaf L of $\mathcal{F}_k$ on $M - V_k$ such that $\bar{L}$ is a compact complex analytic hypersurface (possibly singular) on M while every leaf of $\mathcal{E}_k$ on X is dense. This apparent contradiction (the fact that this is indeed a contradiction needs justification) proves that $F_k = \tilde{M} \to \Delta$ is either holomorphic or anti-holomorphic, hence establishing the strong rigidity theorem for X. Recently, (Mok [11]) we have obtained a new proof of the holomorphicity or anti-holomorphicity of F_k based on the argument that otherwise leaves of $\mathcal{F}_k$ would be contained in level sets S_λ of λ_k (regarded as defined on

$M - V_k$) so that no leaf of $\mathcal{F}_k$ can be dense in M (while every leaf of ${}_k$ on X is dense). This new proof appears to be more adaptable to the study of strong rigidity phenomena for non-compact irreducible quotients of Δ^n of finite invariant volume. Note however that both the existence of harmonic maps and the $\partial\bar{\partial}$-Bochner-Kodaira formula of Siu [18] have not been established in this situation.

The remaining two examples are related to the question of characterizing compact Kähler manifolds of semipositive holomorphic bisectional curvature. Very recently Zhong and the author proved that every compact Kähler-Einstein manifold X of semipositive holomorphic bisectional curvature and positive Ricci curvature is biholomorphically isometric to a compact Hermitian symmetric space (Mok-Zhong [15],[16]). This answers in the affirmative a question raised by Siu [21] in the International Congress of Mathematicians at Warsaw, 1983. The starting point of our proof is the work of Berger [1]. By applying the maximum principle to the curvature function on the unit sphere bundle $S^{1,0}(X)$ of the holomorphic tangent bundle $T^{1,0}(x)$, he showed that a compact Kähler-Einstein manifold X of positive sectional curvature is biholomorphically isometric to the complex projective space $\mathbb{P}^n$ equipped with a Fubini-Study metric, which is characterized by the vanishing of the Kählerian Bochner-Weyl tensor. We obtained our theorem by proving the vanishing of ∇R for the curvature tensor R on X. Our approach, as indicated at the beginning of the Introduction, is to prove some partial vanishing of ∇R at every point of X and then to integrate such information globally. Let $\mathfrak{M} \subset S^{1,0}(X)$ denote the set of all $\alpha \in S^{1,0}(X)$ attaining the global maximum of all holomorphic sectional curvatures. Write $\mathfrak{M}_x = \mathfrak{M} \cap S^{1,0}_x(X)$. Basing in part on Berger's formula we prove that (i) $\mathfrak{M}_x \neq \emptyset$ for any $x \in X$ and (ii) $\nabla_\alpha R = 0$ for a generic $\alpha \in \mathfrak{M}$. For the proof of (ii) we developed a higher order maximum principle for tensors associated to elliptic operators of higher (even) order in place of the Laplacian. Let V_x be the $\mathbb{C}$-linear span of $\mathfrak{M}_x$. Then $\nabla_\xi R = 0$ for $\xi \in V_x$ for a generic $x \in X$. By integrating the distribution $x \to \mathrm{Re}\, V_x$ we obtain totally geodesic compact Hermitian symmetric submanifolds, yielding a de Rham decomposition $X = X_1 \times X_2$ as an isometric product of a compact Hermitian symmetric manifold X_1 and a compact Kahler manifold X_2 satisfying the same hypothesis as X. The proof of our theorem is completed by applying induction on the dimension of X.

Our last example is a very recent result of the author giving a Kähler rigidity theorem on irreducible compact Hermitian symmetric spaces (X,g) of rank ≥ 2 (Mok [12][14], to appear). Namely, if h is any Kähler metric of semipositive holomorphic bisectional curvature on the underlying complex manifold, then (X,h) is already a Hermitian symmetric space. Since there exists a large set of biholomorphic transformations of X not preserving g, our theorem only asserts the uniqueness of h up to biholomorphisms. In view of the above theorem of Mok-Zhong, it suffices to show that (X,h) is Kähler-Einstein. To convert this assertion to some form of vanishing theorem, we introduce the notion of characteristic spheres in X. Namely, fixing g such that (X,g) is a Hermitian symmetric space, a characteristic sphere is a totally geodesic Riemann sphere S such that at each $x \in S$, $T_x^{1,0}(S)$ is generated by some vector $\alpha \in \mathfrak{M}_x$ (definition as in the last paragraph). We observe that the collection of characteristic spheres $\{S\}$ of (X,g) is independent of the choice of g_0. Furthermore, to prove that (X,h) is Kähler-Einstein, it is sufficient to show that every characteristic sphere S is also totally geodesic in (X,h). Thus, we are led to prove the vanishing of certain components of Riemann-Christoffel symbols (Γ^k_{ij}) along characteristic spheres S. Denote locally the Kähler metrics g and h by (g_{ij}) and (h_{ij}) respectively and write $h_{i\bar{j},k}$ for $\nabla_k h_{i\bar{j}}$, the covariant differentiation being performed in terms of $(g_{i\bar{j}})$, the reference metric. We obtain first an integral formula on $\bar{\mathfrak{M}} = \mathfrak{M}/(\text{multiplication by } e^{i\theta}) \subset \mathbb{P}T^{1,0}(X)$ in terms of first Chern classes which yields pointwise the vanishing of certain components of $(h^{i\bar{j}}{}_{,k})$. Then, we apply the Kähler condition and compute along characteristic spheres to obtain the desired vanishing of components of (Γ^k_{ij}), proving our Kähler rigidity theorem.

As is apparent from the discussion of examples above, the

correct choice of foliations or families of submanifolds to globalize pointwise information (obtained from integral formulas or some form of the maximal principle) is vital to the success of the local-global approach to vanishing theorems. In certain situations it dictates the pointwise information needed and thus determines the whole scheme of proof.

We would like to thank Professor Siu for inviting the author to give a talk in Bowdoin College, Maine in August, 1984 on the occasion of the Summer Conference on Nonlinear Differential Equations and Differential Geometry. In addition, we thank Professors Gunning, Jost, Kuranishi, Phong, Siu and Yau for their help and encouragement.

TABLE OF CONTENTS

§1. THE VANISHING THEOREM OF MATSUSHIMA-SHIMURA ON IRREDUCIBLE COMPACT QUOTIENTS OF POLYDISCS.

(1.1) Background

Let Γ be a discrete group of automorphisms of the polydisc Δ^n $(n \geq 2)$ without fixed points such that $X = \Delta^n/\Gamma$ is compact and irreducible (cf. the definition given in the Introduction). Assume first $\Gamma \subset (\mathrm{Aut}(\Delta))^n$. On Δ^n consider the line bundle L_k, $1 \leq k \leq n$ of complexified tangent vectors of the special form $a\frac{\partial}{\partial z_k}$, where

$(z_1,\ldots,z_n)$ is the system of Euclidean coordinates in $\mathbb{C}^n$ and a is an arbitrary complex number. Equip L_k with the Hermitian metric induced by the Poincare metric on Δ^n. Since $\Gamma \subset (\mathrm{Aut}(\Delta))^n$, the Hermitian line bundle is invariant under Γ. We denote the induced Hermitian line bundle on X by $L_k(X)$. For a general $\Gamma \subset \mathrm{Aut}(\Delta^n)$ with X/Γ compact and irreducible consider $\Gamma_0 = \Gamma \cap (\mathrm{Aut}(\Delta))^n$. It is easy to see that Γ_0 is a normal subgroup of finite index. We denote by $\pi: X_0 \to X$ the finite covering of X by $X_0 = \Delta^n/\Gamma_0$ thus obtained. In 1963, Matsushima-Shimura [9] proved

THEOREM (Matsushima-Shimura [9]): Let $X = \Delta^n/\Gamma$ be an irreducible compact quotient of the polydisc $(n \geq 2)$, E a Hermitian holomorphic vector bundle on X such that under the covering map $\pi: X_0 \to X$, π^*E is isomorphic to $L_{j_1}^{s_1}(X_0) \oplus\cdots\oplus L_{j_m}^{s_m}(X_0)$ with $s_k > 0$ for $1 \leq k \leq m \leq n$, $1 < j_1 <\cdots< j_m \leq n$. Then $H^p(X,E) = 0$ for $0 \leq p < n$.

In Matsushima-Shimura [9] the main purpose was to use the Riemann-Roch theorem, Serre duality and their vanishing theorem to compute the dimension of certain spaces of automorphic forms. However, when applied to the bundle E with $\pi^*E = L_1(X_0) \oplus\cdots\oplus L_n(X_0)$, one obtains vanishing theorems for $E = T(X)$, the holomorphic tangent bundle of X. In particular, the vanishing of $H^1(X,T(X))$ signifies the local rigidity of X as a complex manifold, i.e., all differentiably trivial holomorphic families of compact complex manifolds centered at X must be holomorphically trivial. The vanishing theorem of Matsushima-Shimura and its consequences complemented vanishing theorems of Calabi-Vesentini [4] and Borel [2] on compact quotients of irreducible bounded symmetric domains of dimension ≥ 2. The proof of [9] is moreover differential-geometric in nature. We note that a similar theorem in the context of quotients of Lie groups by discrete subgroups was established by Weil [23]. In this article we present a somewhat simplified argument which generalizes immediately to the situation of non-compact irreducible quotients of Δ^n with finite invariant volume (Lai-Mok [8]) by avoiding the use of the maximum principle (cf. [9, p. 428]).

(1.2) A Bochner-Kodaira formula and its consequences.

The starting point of [9] was a Bochner-Kodaira formula. Since the proof of the theorem evidently reduces to the situation of line bundles $L_k^{S_k}(X)$, $X = \Delta^n/\Gamma$ with $\Gamma \subset (\mathrm{Aut}(\Delta))^n$, we formulate the formula in terms of line bundles.

THE BOCHNER-KODAIRA FORMULA FOR THE ∇-GRADIENT: Let X be a compact Kahler manifold, $L \to X$ a Hermitian holomorphic line bundle with the curvature form $\sqrt{-1}/2\ \Sigma_{i,j}\ c_{i\bar{j}} dz_i \wedge d\bar{z}_j$. Let ψ be a smooth L-valued (o,p)-form. In terms of norms and star operators induced by the Kähler metric $(g_{i\bar{j}})$ on X and the Hermitian metric on L, we have

$$\int_X \|\bar{\partial}\psi\|^2 + \|\bar{\partial}^*\psi\|^2$$

$$= \int \|\nabla\psi\|^2 + \frac{1}{(p-1)!}\int_X \Sigma_{i,j,I_{p-1}}\ g^{i\bar{m}}\ c_{i\bar{j}}\psi_{\overline{mI_{p-1}}}\ \overline{\psi^{jI_{p-1}}}$$

$$- \frac{1}{p!}\int_X \Sigma_{i,j,I_p}\ g^{i\bar{j}} c_{i\bar{j}}\psi_{\bar{I}_p}\ \psi^{\bar{I}_p}$$

where I_q stands for the multi-index $(i_1,\ldots,i_q)$ and

$$\psi^{i_1,\ldots,i_p} = \Sigma_{j_1,\ldots,j_p}\ g^{i_1\bar{j}_1} g^{i_2\bar{j}_2} \cdots g^{i_p\bar{j}_p}\ \psi_{\bar{j}_1\ldots\bar{j}_p}.$$

To apply the Bochner-Kodaira formula for the ∇-gradient, let φ now be a harmonic L-valued (o,p) form. It satisfies $\bar{\partial}\varphi = \bar{\partial}^*\varphi = 0$. We can write

$$\int_X \|\nabla\varphi\|^2 + K_p(\varphi,\varphi) = 0$$

where K_p is a Hermitian bilinear form at each point $x \in X$ on $L_x \otimes (\Lambda^p T_x^{1,0}(X))$ such that $K_p(\varphi,\varphi) \geq 0$ whenever L carries seminegative curvature. When L carries strictly negative curvature and $0 \leq p \leq n = \dim_{\mathbb{C}} X$, K_p is pointwise a positive definite Hermitian bilinear form from which one concludes immediately the vanishing of φ. This is the Kodaira vanishing theorem for negative line bundles on Kähler manifolds.

In our present situation with $X = \Delta^n/\Gamma$ compact irreducible, $\Gamma \subset (\mathrm{Aur}(\Delta))^n$ and $L = L_k^s(X)$, $s > 0$, $1 \leq k \leq n$ $(n \geq 2)$, L carries only seminegative curvature in the Poincaré metric. In terms of Euclidean coordinates on the polydisc, we have

$$K_p(\varphi,\varphi) = \frac{1}{(p-1)!} g^{k\bar{k}} c_{k\bar{k}} \Sigma_{I_{p-1}} \|\varphi_{\bar{k}\bar{I}_{p-1}}\|^2 - \frac{1}{p!} g^{k\bar{k}} c_{k\bar{k}} \Sigma_{I_p} \|\varphi_{\bar{I}_p}\|^2$$

$$= -\frac{1}{p!} g^{k\bar{k}} c_{k\bar{k}} \Sigma_{k\notin I_p} \|\varphi_{\bar{I}_p}\|^2 \geq 0 \text{ since } c_{k\bar{k}} < 0.$$

Thus, one can make use of the Bochner-Kodaira formula and the explicit formula for $K_p(\varphi,\varphi)$ to deduce that $\nabla\varphi = 0$ and $\varphi_{\bar{I}_p} = 0$ whenever $k \notin I_p$. Write $\tilde{\varphi}$ for the lift of φ to Δ^n and let $\tilde{\varphi} = \Sigma_{i_1<\ldots<i_p} \tilde{\varphi}(i_1,\ldots,i_q)$ be the obvious decomposition with $\tilde{\varphi}(i_1,\ldots,i_p) = \tilde{\varphi}_{i_1,\ldots,i_p}(\frac{\partial}{\partial z_p})^{s_k} \otimes (dz_{i_1} \wedge\ldots\wedge dz_{i_p})$. Since $\Gamma \subset (\mathrm{Aut}(\Delta))^n$ the decomposition of $\tilde{\varphi}$ induces a decomposition $\varphi = \Sigma\varphi(i_1,\ldots,i_p)$. Moreover, $\varphi(i_1,\ldots,i_p)$ is harmonic since the Kähler Laplacian $\square = \bar{\partial}\bar{\partial}^* + \bar{\partial}^*\bar{\partial}$ acts componentwise on Δ^n. Thus we obtain

$$\begin{cases} \bar{\partial}\varphi(i_1,\ldots,i_p) = \bar{\partial}^*\varphi(i_1,\ldots,i_p) = 0 \\ \nabla\varphi(i_1,\ldots,i_p) = 0. \end{cases}$$

From these one easily deduces

$$\tilde{\varphi}_{I_p} = \tilde{\varphi}_{I_p}(z_{i_1}, z_{i_2}, \ldots, z_{i_p})$$

i.e. it depends only on $(z_{i_1},\ldots,z_{i_p})$. Recall also from the Bochner-Kodaira formula that $\varphi_{I_p} = 0$ whenever $k \notin I_p$. Thus we are led to study $L_k^s(X)$-valued harmonic (0,p) forms φ of the particular form ($\tilde{\varphi}$ denoting the lifting of φ to Δ^n)

$$\tilde{\varphi} = f(z_1,\ldots,z_p)(\frac{\partial}{\partial z_1})^s \otimes (dz_1 \wedge\ldots\wedge dz_p)$$

by taking $k = 1$ and $I_p = (1,\ldots,p)$. Finally the vanishing of such a φ will be obtained by global considerations.

(1.3) Global Considerations.

Consider now the holomorphic foliation $\tilde{\mathcal{E}}$ on Δ^n whose leaves are given by $\{z'\} \times \Delta^{n-p}$, $z' \in \Delta^p$ where p satisfies $0 < p < n$. (For $p = 0$ the vanishing of $H^0(X, L_1^s(X))$ for $s > 0$ is trivial.) Let $\mathcal{E}$ be the induced foliation on $X = \Delta^n/\Gamma$, with $\Gamma \subset (\mathrm{Aut}(\Delta))^n$ and X is compact and irreducible. Let now φ be a harmonic $L^s(X)$-valued (0,p) form such that the lifting $\tilde{\varphi}$ to Δ^n is given by

$\tilde{\varphi} = f(z_1,\ldots,z_p)(\frac{\partial}{\partial z_1})^s \otimes (dz_1 \wedge\ldots\wedge dz_p)$. Recall that the Hermitian metric on L_1^s on Δ^n is defined by

$$\|(\frac{\partial}{\partial z_1})^s\|(z_1,\ldots,z_n) = \frac{1}{(1-|z_1|^2)^s}.$$

Thus,

$$\|(\frac{\partial}{\partial z_1})^s \otimes (d\bar{z}_1 \wedge\ldots\wedge d\bar{z}_p)\|(z_1,\ldots,z_n)$$

$$= \frac{1}{(1-|z_1|^2)^{s-1}}(1-|z_2|^2)^2 \ldots (1-|z_p|^2)$$

is independent of $(z_{p+1},\ldots,z_n)$. In particular, the length of $\tilde{\varphi}$ on each leaf $\{z\} \times \Delta^{n-1}$ of $\mathcal{E}$ is constant. To study the behavior of φ using $\mathcal{E}$ we have the following lemma.

LEMMA (Matsuchima-Shimura [9], using Borel [3]): Let $\Gamma \subset (\mathrm{Aut}(\Delta))^n$ be a discrete subgroup such that $X = \Delta^n/\Gamma$ is compact and irreducible. Then, the projection of Γ onto any direct factor $(\mathrm{Aut}(\Delta))^k$, $1 \leq k < n$ has dense image.

It follows immediately from the lemma that each leaf of the holomorphic foliation $\mathcal{E}$ is dense in X. Since the length of $\tilde{\varphi}$ is constant on each leaf of $\tilde{\mathcal{E}}$ it follows that $\|\varphi\|$ = constant on X. We assert this leads to a contradiction unless $\varphi \equiv 0$. Consider the $\bar{L}_1^{s}$-valued $(p,0)$ form $\bar{\varphi}$ (pointwise a multiple of $(\frac{\partial}{\partial \bar{z}_1})^s \otimes (dz_1 \wedge\ldots\wedge dz_p)$). Let χ be obtained from $\bar{\varphi}$ by contracting with $(g^{1\bar{1}})^s$ in the obvious way to get a section of $L_1^{-s} \otimes (L_1^{-1} \otimes\ldots\otimes L_p^{-1})$. χ is pointwise a multiple of $dz_1^s \otimes (dz_1 \wedge\ldots\wedge dz_p)$. Since $\nabla\varphi = 0$, $\bar{\nabla}\chi = 0$. It follows that χ is holomorphic. But clearly $\|\chi\| = \|\varphi\|$ so that $\Delta \log \|\chi\| = 0$ while in general $\Delta \log\|\chi\|$ is the curvature form of $L_1^{-s} \otimes (L_1^{-1} \otimes\ldots\otimes L_p^{-1})$ whenever $\chi \neq 0$. Thus we have a contradiction unless φ vanishes identically on X, proving Theorem 1.

§2. THE STRONG RIGIDITY OF IRREDUCIBLE COMPACT QUOTIENTS OF POLYDISCS.

(2.1) Background and the $\partial\bar{\partial}$-Bochner-Kodaira formula of Siu.

In 1980, Siu [18] proved the strong rigidity theorem of compact quotients of irreducible classical bounded symmetric domains. This result was further extended to cover the situation of exceptional bounded symmetric domains in Siu [19]. The results can be stated as follows.

THEOREM (Siu [18][19]): Let $X = \Omega/\Gamma$ be a compact quotient of an irreducible bounded symmetric domain Ω of dimension ≥ 2. Let M be a compact Kahler manifold diffeomorphic to X. Then there exists a diffeomorphism $f: M \to X$ which is either holomorpic or anti-holomorphic.

If one assume in addition that M is locally symmetric, the result follows from a special case of Mostow's rigidity theorem. When M is only assumed to be Kähler, the proof of Siu's strong rigidity theorem is based on his $\partial\bar{\partial}$-Bochner-Kodaira formula for smooth mappings between Kahler manifolds. For any smooth mapping $f: M \to N$ between two complex manifolds the induced mapping $df = T(M) \to T(N)$ between the real tangent bundles decomposed after complexification into components ∂f, $\bar{\partial} f$, $\partial \bar{f}$, $\bar{\partial}\bar{f}$, where $\bar{\partial} f$, for example, maps $T^{0,1}(M)$ into $T^{1,0}(N)$. Such a decomposition is obtained from the decomposition $T(X) \otimes_R \mathbb{C} = T^{1,0}(X) \oplus T^{0,1}(X)$ for any complex manifold X. One can also interpret $\bar{\partial} f: T^{0,1}(M) \to T^{1,0}(N)$ as an $f^*T^{1,0}(N)$-valued $(0,1)$ form on M. f is holomorphic (anti-holomorphic) if and only if $\bar{\partial} f = 0$ ($\partial f = 0$). For any section s of $f^*T^{1,0}(N)$ on some open subset U of M, one can define covariant derivatives

$$\frac{Ds^\alpha}{\partial z_i}(P) = \frac{\partial s^\alpha}{\partial z_i}(P) + \sum_{\beta,\gamma} {}^N\Gamma^\alpha_{\beta\gamma}(f(P))s^\beta \frac{\partial f^\gamma}{\partial z_i}(P)$$

$$\frac{Ds^\alpha}{\partial \bar{z}_i}(P) = \frac{\partial s^\alpha}{\partial \bar{z}_i}(P) + \sum_{\beta,\gamma} {}^N\Gamma^\alpha_{\beta\gamma}(f(P))s^\beta \frac{\partial f^\gamma}{\partial \bar{z}_i}(P)$$

where $P \in U$ and (z_i) is a system of local holomorphic coordinates on U. Denote locally by $g = 2 \operatorname{Re} \Sigma\, g_{\alpha\bar{\beta}}\, dz^\alpha \otimes d\bar{z}^\beta$ the Kähler metric on N and by $R^N = \Sigma\, R_{\alpha\bar{\beta}\gamma\bar{\delta}}\, dz^\alpha \otimes d\bar{z}^\beta \otimes dz^\gamma \otimes d\bar{z}^\delta$ the curvature tensor on N regarded as a $(2,2)$ tensor with the convention chosen such that $R_{\alpha\bar{\alpha}\gamma\bar{\gamma}}$ would be negative for the unit ball equipped with the Poincaré metric. With these notations we have

THE $\partial\bar{\partial}$-BOCHNER-KODAIRA FORMULA OF SIU [18]: Let M and N be Kähler manifolds and $f: M \to N$ be a smooth mapping. Denote by

$\langle\ ,\ \rangle_N$ norms induced by the Kahler metric on N. Then, on M

$$\partial\bar{\partial}\langle g, \bar{\partial}f \wedge \partial\bar{f}\rangle_N = \langle R^N, \bar{\partial}f \wedge \partial\bar{f} \wedge \partial f \wedge \overline{\partial\bar{f}}\rangle_N - \langle g, D\bar{\partial}f \wedge \bar{D}\partial\bar{f}\rangle_N.$$

To make use of the above formula to prove strong rigidity theorems, Siu used an integrated form of the formula for harmonic maps. Let ω_M be the Kahler form of M. Then, from Stokes' Theorem and the formula $-\langle g, D\bar{\partial}f \wedge \bar{D}\partial\bar{f}\rangle_N \wedge \omega_M^{n-2} = \|D\bar{\partial}f\|^2$ for harmonic maps we have

$$\int_M \|D\bar{\partial}f\|^2 + \langle R^N, \bar{\partial}f \wedge \partial\bar{f} \wedge \partial f \wedge \overline{\partial\bar{f}}\rangle \wedge \omega_M^{n-2} = 0.$$

In this form the analogy with the Bochner-Kodaira formula in §1 is apparent. Given $X = \Omega/\Gamma$ as in the strong rigidity theorem of Siu and $f_0: M \to X$ a diffeomorphism. From the existence theory of harmonic maps f_0 is homotopic to some harmonic map $f: M \to X$, since X carries seminegative Riemannian sectional curvature. From the preceding integral formula one obtains the vanishing of both the gradient term $D\bar{\partial}f$ and the curvature term. The proof of the vanishing of $\bar{\partial}f$ or ∂f was obtained from algebraic manipulations coming from the vanishing of the curvature term. We note that to make use of the integral formula one has to show that the curvature tensor of a bounded symmetric domain (equipped with the Kähler-Einstein metric) verifies a stronger notion of seminegativity (Siu [ ][ ]). In fact, the Hermitian biliniar form Q on $T_x^{1,0} \otimes T_x^{0,1}$ at a point x defined by $Q(\xi \otimes \bar{\xi}', \zeta \otimes \bar{\zeta}') = R(\xi, \bar{\xi}'; \zeta', \bar{\zeta})$ (and extended by Hermitian bilinearity) is negative semi-definite.

(2.2) Foliations associated with harmonic maps into compact quotients of polydiscs.

Siu's strong rigidity theorem and its method does not cover the situation of irreducible compact quotients of polydiscs. The strong rigidity of such quotients $X = \Delta^n/\Gamma$ was proved independently by Jost-Yau [6][7] for $n = 2$ and by Mok [10] for all $n \geq 2$. Combined with Siu's theorem, it was further proved in Mok [10] that strong rigidity holds for a compact quotient $X = \Omega/\Gamma$ of a bounded symmetric domain Ω whenever Mostow's rigidity theorem holds for X.

The starting point is again the integrated form of the $\partial\bar{\partial}$-Bochner-Kodaira formula of Siu [18] for harmonic maps. In the present situation the vanishing of the curvature term itself is not strong enough to prove the vanishing of $\bar{\partial}f$ or ∂f. However, by combining the vanishing of $D\bar{\partial}f$ (i.e. f is a pluriharmonic map) and the vanishing of the curvature term (compare this with §1!) Jost-Yau proved that f gives rise to holomorphic foliations defined on dense open subsets of M. More precisely, we have

PROPOSITION (Jost-Yau [6]): Let $f = M \to X = \Delta^n/\Gamma$ be a harmonic map from a compact Kähler manifold M into a compact quotient of the unit polydisc, equipped with the standard Poincaré metric. Let $F = (F_1,\ldots,F_n): \tilde{M} \to \tilde{X} = \Delta^n$ be the lifting of f to the universal coverings. Suppose $\tilde{P} \in \tilde{M}$ is such that F_i is of real rank 2 at $\tilde{P}$. Then, there exists a system of local holomorphic coordinates $(z;z')$, $z = z_1$, $z' = (z_2,\ldots,z_n)$, $z \in \Delta$, $z' \in \Delta^{n-1}$ on an open neighborhood $\tilde{U}$ of $\tilde{P}$ such that the mapping $F_i: \tilde{U} \to \Delta$ is of the form $F_i(z,z') = g_i(z)$.

Remarks.
We shall call the system of local coordinates (z,z') on $\tilde{U}$ as above privileged with respect to F_i.

PROOF OF PROPOSITION: From the integrated $\partial\bar{\partial}$-Bochner-Kodaira formula of Siu we have

$$\begin{cases} D\bar{\partial}F_i = 0 \\ \langle R, \bar{\partial}F_i \wedge \partial\bar{F}_i \wedge \partial F_i \wedge \bar{\partial}\bar{F}_i\rangle = 0 \end{cases}$$

everywhere on $\tilde{M}$, where $R = R^{\Delta^n}$ is the curvature tensor of the Poincare metric on Δ^n. In terms of Euclidean coordinates we have $R_{\alpha\bar{\alpha}\alpha\bar{\alpha}} < 0$ for $1 \leq \alpha \leq n$ and all other terms are zero. It follows that the vanishing of the curvature term implies $\bar{\partial}F_i \wedge \bar{\partial}\bar{F}_i \equiv 0$ on $\tilde{M}$. In terms of Riemann-Christoffel symbols on Δ^n, the equation $D\bar{\partial}F_i = 0$ means that for local holomorphic coordinates $(z_1,\ldots,z_n)$ on M, we have

$$(*) \qquad \frac{\partial^2 F_i}{\partial z_\alpha \partial \bar{z}_\beta} + \Gamma^i_{ii}\frac{\partial F_i}{\partial z_\alpha}\frac{\partial F_i}{\partial \bar{z}_\beta} = 0$$

where it is understood that Γ^i_{ii} means $\Gamma^i_{ii}(F_i(\tilde{P}))$ when other terms are evaluated at $\tilde{P} \in \tilde{M}$. Let $\tilde{P}$ be such that real rank $F_i(\tilde{P}) = 2$.

In a neighborhood of $\widetilde{P}$ level sets of F_i define a real analytic foliation $\widetilde{\mathcal{F}}_1$ with closed real 2-codimensional fibers. ($f: M \to X$ is real-analytic because it is a pluriharmonic map into a complex manifold equipped with a real-analytic metric.) The proposition is equivalent to asserting that this foliation is holomorphic. First, to see that the leaves are holomorphic it suffices to show that the (a priori integrable) distribution $\widetilde{x} \to \ker dF_i(\widetilde{x})$ in a neighborhood $\widetilde{U}$ of $\widetilde{P}$ is closed under the J-operator. Assuming that F_i is of real rank 2 on $\widetilde{U}$, we have, in local holomorphic coordinates $(z_1,\ldots,z_n)$ on $\widetilde{U}$, $\partial F_i/\partial \bar{z}_j = \lambda\, \partial \bar{F}_i/\partial \bar{z}_j$ (since $\bar{\partial} F_i \wedge \bar{\partial}\bar{F}_i = 0$) with $|\lambda| \neq 1$. Assume without loss of generality that $\partial F_i \neq 0$ on $\widetilde{U}$. It is then easy to check that, for a real tangent vector X on $\widetilde{U}$ at $\widetilde{x}$, $\langle dF_i(\widetilde{x}), X\rangle = 0$ if and only if $\langle \partial F_i(\widetilde{x}), X_{1,0}\rangle = 0$ where $X = X_{1,0} + X_{0,1}$ is the decomposition of X into (1,0) and (0,1) components. This uses the fact that $|\lambda| \neq 1$. Thus $\ker dF_i(\widetilde{x})$ corresponds to the $\mathbb{C}$-linear space $\ker \partial F_i(\widetilde{x})$ (acting on $T^{1,0}_{\widetilde{x}}(\widetilde{M})$) under the correspondence $X \mapsto \frac{1}{2}(X - \sqrt{-1}JX)$. This forces $\ker dF_i(\widetilde{x})$ to be invariant under J so that leaves of $\mathcal{F}_i$ on $\widetilde{U}$ are holomorphic. To show that $\mathcal{F}_i$ is in fact a holomorphic foliation on $\widetilde{U}$ it suffices to show that the map $\widetilde{x} \mapsto \ker \partial F_i(\widetilde{x})$ is holomorphic. Denote by $[\omega]$ the element of $\mathbb{P}T^{1,0}_{\widetilde{x}}(\widetilde{M})$ defined by $\omega \in T^{1,0}_{\widetilde{x}}(\widetilde{M})$. Then, $\widetilde{x} \mapsto \ker \partial F_i(\widetilde{x})$ is holomorphic if and only if $\widetilde{x} \mapsto [\partial F_i(\widetilde{x})]$ is holomorphic. Assuming $\frac{\partial F_i}{\partial z_1} \neq 0$ on $\widetilde{U}$, it suffices to show that $g_\alpha = \frac{\partial F_i}{\partial z_\alpha}\Big/\frac{\partial F_i}{\partial z_1}$ is holomorphic. By the homogeneity of the Poincaré disc we may assume without loss of generality that $F_i(\widetilde{x}) = 0$, where $\Gamma^i_{ii} = 0$. Then (*) simply means

$$\frac{\partial^2 F_i}{\partial z_\alpha \partial \bar{z}_\beta}(\widetilde{x}) = 0 \text{ for } 1 \leq \alpha,\beta \leq n. \quad \text{Thus,}$$

$$\begin{aligned}
\frac{\partial g_\alpha}{\partial \bar{z}_\beta}(\widetilde{x}) &= \frac{\partial}{\partial \bar{z}_\beta}\left(\frac{\partial F_i}{\partial z_\alpha}\Big/\frac{\partial F_i}{\partial z_1}\right)(\widetilde{x}) \\
&= \left[\left(\frac{\partial F_i}{\partial z_1}\frac{\partial^2 F}{\partial z_\alpha \partial \bar{z}_\beta} - \frac{\partial F_i}{\partial z_\alpha}\frac{\partial^2 F_i}{\partial z_1 \partial \bar{z}_\beta}\right)\Big/\left(\frac{\partial F_i}{\partial z_1}\right)^2\right](\widetilde{x}) \\
&= 0
\end{aligned}$$

proving the proposition.

In Jost-Yau [6][7] conditions were found so that the mapping f is a diffeomorphism so that using the holomorphic foliations $\tilde{\mathcal{F}}_i$ and Mostow's rigidity theorem, the strong rigidity of irreducible compact quotinets of polydiscs was proved for $n = 2$. In [10] we studied instead the induced foliation $\mathcal{F}_i$ on M and proved

THEOREM (Mok [10]): Let M be a compact Kähler manifold and X be an irreducible compact quotient of Δ^n, $n \geq 2$. Write $F = (F_1,\ldots,F_n)$ for the lifting $F: \tilde{M} \to \Delta^n$ to the universal coverings. Suppose real rank $F \geq 3$ and real rank $F_i = 2$. Then F_i is either holomorphic or anti-holomorphic.

Here by the real rank we mean the maximal rank of the differential on M. The proof of our theorem (which is stronger than a strong rigidity theorem) is based on comparing the dynamical behavior of the foliations $\mathcal{F}_i$ on dense open subsets of M and $\mathcal{E}_i$ on X, defined by projecting $\Delta^{n-1} \times \{z_i\} \times \Delta^{n-i}$ into X.

(2.3) <u>Dynamical behavior of $\mathcal{F}_i$ on M.</u>

Let $\tilde{Y}_i \subset \tilde{M}$ be the real-analytic subvariety on which F_i fails to be a submersion. $\tilde{Y}_i$ is invariant under $\pi_1(M)$, so that it corresponds to a real-analytic subvariety Y_i on M. The holomorphic foliation is a priori only defined on $M - Y_i$, which may be disconnected. The first step of our proof is to extend $\mathcal{F}_i$ to some connected open subset of M. The proof will be achieved by an argument by contradiction. Roughly, if F_i were neither holomorphic nor anti-holomorphic, the (extended) foliation $\mathcal{F}_i$ would have a closed leaf while every leaf of the corresponding foliation $\mathcal{E}_i$ on X is dense. To make this argument rigorous, the proof is broken up into the following steps. (As before we assume $\Gamma \subset (\mathrm{Aut}(\Delta))^n$ without loss of generality.)

I) There exists a complex-analytic subvariety V_i of M of complex codimension ≥ 2, $V_i \subset Y_i$, such that the holomorphic foliation $\mathcal{F}_i$ extends holomorphically from $M\text{-}Y_i$ to $M\text{-}V_i$.

II) If F_i were neither holomorphic nor anti-holomorphic, there would exist a closed leaf L of the extended foliation $\mathcal{F}_i$ on $M\text{-}V_i$ such that $\bar{L}$ is a (compact) complex-analytic subvariety of M.

III) Set $i = 1$. Then, $f_*(\pi_1(\bar{L}))$ would be a non-trivial subgroup of $\{id_\Delta\} \times (\mathrm{Aut}(\Delta))^{n-1}$. As a consequence $\Phi = \Gamma \cap \{id_\Delta\} \times (\mathrm{Aut}(\Delta))^{n-1}$ is a non-trivial normal subgroup of Γ.

IV) The existence of such a Φ contradicts the lemma of Borel-Matsushima-Shimura (cf. (1.3)).

We give here a brief justification of the proof. I) To extend $\mathcal{F}_i$ it suffices to extend the holomorphic mapping $\tilde{x} \longmapsto [\partial F_i(\tilde{x})] \in \mathbb{P}T^{1,0}_{\tilde{x}}(\tilde{M})$ on $\tilde{M} - \tilde{Y}_i$ (assuming $\partial F_i \not\equiv 0$ on $\tilde{M}$). The question being local, it is equivalent to extend meromorphically to B^n a holomorphic function h on B^n—(some real-analytic subvariety S) which is at the same time the quotient of two real-analytic functions on B^n $(h = u/v)$. By expanding in Taylor series both u and v about a generic point of S it is easy to see that there exists a real-analytic subvariety S' of S, $\dim_{\mathbb{R}} S' \leq 2n - 2$ such that h extends holomorphically to $B^n - S'$. We may assume S' irreducible. If S' is a complex analytic subvariety, then h extends meromorphically across S' because h cannot have essential singularities along S' (recall $h = u/v$ on B^n). If S' is not complex analytic then at generic points of S' h extends even holomorphically by the Theorem of Oka (on characterizing domains of holomorphy). By the Theorem of Hartogs h extends holomorphically to B^n in this situation.

II) Let $\lambda = \left|\frac{\partial \bar{F}_i/\partial z}{\partial F_i/\partial z}\right|$ in terms of a local system of privileged coordinates (z,z') adapted to $\mathcal{F}_i$. Then λ is invariant under $\pi_1(M)$. Thus, it can also be regarded as a function on $M - Y_i$. We have $\Delta \log \lambda = 2(\|\bar{\partial} F_i\|^2 - \|\partial F_i\|^2) = 2\|\partial F_i\|^2 (\lambda^2 - 1)$ whenever $\lambda \neq 0$. From this and elementary potential-theoretic considerations λ can be extended as a continuous function from $M - V_i$ to $[0,\infty]$. If λ attains both the supremum and the infimum on $M - V_i$ it is easy to prove from the above formula and the maximum principle that λ must have some zero or some pole on $M - V_i$. Using the fact that V_i is of complex codimension ≥ 2 and that λ is constant along leaves of $\mathcal{F}_i$ one can indeed show that the infimum and the supremum of λ are attained. Let now L be a connected component of either the zero set or the pole set of λ on $M - V_i$. Again from potential-theoretic considerations one can show that L is of real codimension 2 in

M, thus identifying with some leaf of $\mathcal{F}_i$. Finally, $\bar{L}$ is a (compact) complex analytic subvariety of M by the extension theorem of Remmert-Stein.

III) That $f_*(\pi_1(\bar{L})) \neq \{id\}$ is obtained by a repeated application of the maximum principle of harmonic maps into negatively-curved Riemannian manifolds $f_*(\pi_1(\bar{L})) \subset \{id_\Delta\} \times (\mathrm{Aut}(\Delta))^{n-1}$ essentially because for any $\tilde{L}$ lying in M above L, $F(\tilde{L}) \subset \text{point} \times \Delta^{n-1}$. Obviously $\Phi = \Gamma \cap \{id_\Delta\} \times (\mathrm{Aut}(\Delta))^{n-1}$ is nontrivial and normal in Γ.

IV) Φ is discrete in $\{id_\Delta\} \times (\mathrm{Aut}(\Delta))^{n-1}$. Let $\varphi \in \Phi$ be nontrivial, $\varphi = (id, \varphi')$. For any $\gamma \in \Gamma$ sufficiently close to the identity, we conclude from $\gamma\varphi\gamma^{-1} \in \Phi$ that in fact $\gamma\varphi = \varphi\gamma$. Write $\gamma = (\gamma_1, \gamma')$. It follows from the above that γ' centralizes φ'. Let Γ' be the projection of Γ into $(\mathrm{Aut}(\Delta))^{n-1}$ given by $\gamma \mapsto \gamma'$. Then the identity component of $\bar{\Gamma}'$ must be contained in the centralizer G of φ' in $(\mathrm{Aut}(\Delta))^{n-1}$, contradicting $\bar{\Gamma}' = (\mathrm{Aut}(\Delta))^{n-1}$ from the lemma of Borel-Matsushima-Shimura.

In Mok [11] we have another method of proving the strong rigidity theorem for $X = \Delta^n/\Gamma$. Suppose F_i is neither holomorphic nor anti-holomorphic and λ is as defined above. Then, generic level sets S_λ of λ on $M - V_i$ are of real codimension 1 and every leaf of $\mathcal{F}_i$ must be contained in some S_λ (so that no leaf of $\mathcal{F}_i$ could be dense in M). In Mok [11], under a different hypothesis, we showed that the existence of S_λ saturated for the foliation $\mathcal{F}_i$ already leads to a contradiction. The proof yields the following theorem which neither includes nor subsumes the preceding theorem.

THEOREM (Mok [11]): Suppose $f: M \to X = \Delta^n/\Gamma$ is a harmonic map from a compact Kähler manifold into an irreducible compact quotient of the polydisc $(n \geq 2)$ such that $f_*(\pi_1(M))$ is of finite index in $\Gamma = \pi_1(X)$. Then, for the lifting $F = (F_1, \ldots, F_n) = \tilde{M} \to \Delta^n$ to the universal coverings, F_i is either holomorphic or anti-holomorphic.

SKETCH OF PROOF: Suppose all leaves L of $\mathcal{F}_i$ in $M - V_i$ were closed submanifolds. Then $\bar{L} \subset M$ is complex analytic. Let $\varphi_i: M \to \mathfrak{D}$ be the associated mapping to the Douady space of M. Then, the mapping $f: M \to X$ can be factored through $(\varphi_1, \ldots, \varphi_n): M \to \varphi_1(M) \times \ldots \times \varphi_n(M)$, each $\varpi_i(M)$ being a possibly singular compact Riemann surface. This leads to a contradiction by considering fundamental groups. Suppose say, $\mathcal{F}_1$ has a leaf L which is not closed. Suppose $\lambda = a$ on L and S_a is a connected component of the level set $\{\lambda = a\}$ of $\mathcal{F}_1$. S_a is of real codimension 1.

Denote by $pr_1: \Delta^n \to \Delta$ the canonical projection into the first factor. By the lemma of Borel-Matsushima-Shimura, to prove our theorem it is sufficient to show that $pr_1[f_*(\pi_1(M))(w)]$ is not dense in Δ for some $w \in \Delta^n$. This will now be proved by examining the curves $F_1(\tilde{S}_\lambda)$ traced by connected components of level sets of λ on $\tilde{M} - \tilde{V}_1$ under the map F_1. Let $x \in L$ be a point such that L is not a closed submanifold in any neighborhood of x. Let $\tilde{f}: U \to \Delta^n$ be a local lifting of $f: U \to X$ for a sufficiently small neighborhood of x. Generically, $\tilde{f}_1(S_\alpha \cap U)$ is a real-analytic curve C on $\Delta(\tilde{f} = (\tilde{f}_1, \ldots, \tilde{f}_n))$, while $\tilde{f}(L \cap U)$ is a sequence of points $\{b_\nu\}$ on C having $\tilde{f}_1(x) = b_0$ as a point of accumulation. It is easy to see that $b_\nu = \gamma_1^{(\nu)}(b_0)$ for some $\gamma_1^{(\nu)} \in Aut(\Delta)$ such that $(\gamma_1^{(\nu)}, \gamma'^{(\nu)}) \in f_*(\pi_1(M))$ for some $\gamma'(\nu) \in (Aut(\Delta))^{n-1}$. From $\gamma_1^{(\nu)}$ we were able to extract a "generator" γ, such that all b_ν lies on $G(b_\nu)$ for a one parameter subgroup G of $Aut(\Delta)$ containing γ_1. Finally, it is easy to conclude from this that for any $w \in \Delta^n$, $pr_1[f_*(\pi_1(M))(w)]$ lies inside just one orbit $G(w)$ of G. Since $G(u)$ is necessarily a closed curve in Δ this proved our assertion and hence the theorem.

In Siu [22] a more intrinsic method of defining the foliations $\mathcal{F}_i$ was devised. For example, if $f: M \to R$ is a harmonic map from a compact Kähler manifold into a compact negatively-curved Riemann surface. Then he showed that $f^*T^{1,0}(R)$ can be endowed the structure of a holomorphic line bundle such that ∂f is a holomorphic section of $f^*T^{1,0}(R) \otimes \Omega_M^1$, the latter being the cotangent bundle of M. The foliation $\mathcal{F}$ on M is now defined by the distribution $x \longmapsto \kappa er\ \partial f(x)$ (acting on $T_x^{1,0}(M)$). The holomorphicity of the foliation $\mathcal{F}$ outside the zero set of ∂f and its extension to $M - V$ for some complex-analytic subvariety V of codimension ≥ 2 follow rather trivially.

§3. CHARACTERIZING COMPACT HERMITIAN SYMMETRIC SPACES AMONG KAHLER-EINSTEIN MANIFOLDS BY CURVATURE CONDITIONS.

(3.1) The curvature tensor of compact Hermitian symmetric spaces.

In order to motivate part of the discussion of both §3 and §4, we begin by briefly stating properties of the curvature tensor of irreducible compact Hermitian symmetric spaces (cf. Borel [2], Siu [19] and Zhong [21]).

Let G be a real, compact, simple and simply-connected Lie group with trivial center. Let K be the identity component of some involutive automorphism σ of G, such that the center of K is one-dimensional. Let g, k be the Lie albebras of G and K, and p be the orthogonal complement of k in g with respect to the Killing form $B(\cdot,\cdot)$ of g. The $B(\cdot,\cdot)$ is negative definite on g. Identifying p with the real tangent space of G/K at the origin $o = eK$, we can define in a natural way a Riemannian metric on G/K which is induced by $-B$ at o. G/K carries a natural complex structure whose J-structure at the origin is defined by $\mathrm{ad}z\colon p \to p$, $J(v) = [z,v]$, for some non-trivial central element z of k. Write X for G/K. With respect to the given complex structure the Riemannian metric on $(X,T(X))$ defined by the Killing form corresponds to a Hermitian metric ω on $(X,T^{1,0}(X))$. Let $p_{\mathbb{C}} = p \otimes_{\mathbb{R}} \mathbb{C} = p^+ \oplus p^-$ be the orthogonal decomposition of $p_{\mathbb{C}}$ into eigenspaces of the J-operator corresponding to eigenvalues $\sqrt{-1}$ and $-\sqrt{-1}$. p^+ can be identified with $T_0^{1,0}(X)$ and p^- with $T_0^{0,1}(X)$. By homogeneity of (X,ω) for the purpose of computing curvatures one can consider simply the curvature tensor at o.

The Hermitian metric ω on X is Kähler. Denote by R the curvature tensor on X at the origin. Then, for $\xi,\eta \in p^+ = T_0^{1,0}(X)$, we have

$$R_{\xi\bar{\xi}\eta\bar{\eta}} = \|[\xi,\bar{\eta}]\|^2$$

in terms of the Lie bracket on $g = k + p$ extended naturally to $g_{\mathbb{C}} = g \otimes_{\mathbb{R}} \mathbb{C}$. In particular, (X,ω) carries semipositive holomorphic bisectional curvature. For $\xi,\eta \in p^+$, let $\xi\circ\eta \in S^2p^{-1}$ denote their symmetric tensor product. Associated with the curvature tensor R are two Hermitian bilinear operators $P: S^2p^+ \times S^2p^+ \to \mathbb{R}$ and $Q: (p^+ \otimes p^-) \times (p^+ \otimes p^-) \to \mathbb{R}$ defined respectively by

$$P(\xi\circ\eta;\xi'\circ\eta') = R_{\xi\bar{\xi}'\eta\bar{\eta}'}$$

$$Q(\xi\otimes\bar{\xi}';\eta\otimes\bar{\eta}') = R_{\xi\bar{\xi}'\eta\bar{\eta}'}$$

and extended by Hermitian bilinearity. The operators P and Q enjoy a number of universal properties for all $X = G/K$. We state the useful ones for us.

(A) The operator Q is positive semidefinite.

(B) Let α be a dominant weight vector of the K-representation space p^+ over $\mathbb{C}$. Then S^2p^+ decomposes as an orthogonal direct sum $V_1 \oplus V_2$ of irreducible K-representation spaces consisting of eigenspaces of the operator P such that $\alpha\circ\alpha$ is a dominant weight vector of the K-representation space V_1.

The property (A) comes from the explicit expression

$$\begin{aligned} Q(\Sigma\, a_{i\bar{j}}\xi_i \otimes \bar{\eta}_j, \Sigma\, a_{i\bar{j}}\xi_i \otimes \bar{\eta}_j) \\ = \|\Sigma\, a_{i\bar{j}}[\xi_i, \bar{\eta}_j]\|^2. \end{aligned}$$

In the terminology of Hermitian geometry, we say that the curvature operator is semipositive in the dual sense of Nakano (cf. Siu [20]). It implies that X carries seminegative Riemannian sectional curvature. Property (B) is proved in Calabi-Vesentini [4] and Borel [2]. In §3 and §4, the space of unit tangent vectors of type (1,0) attaining the maximum holomorphic sectional curvature plays an important role. Property (B) becomes meaningful to us in view of

(C) $\alpha \in p^+$, $\|\alpha\| = 1$ attains the maximum holomorphic sectional curvature at o if and only if α is a dominant weight vector of the K-representation space p^+ over $\mathbb{C}$ (relative to some Cartan subalgebra of k).

The proof of Property (C) can be found in Mok [14].

A further well-known property for irreducible compact Hermitian symmetric spaces is:

(D) (X,ω) is Kähler-Einstein with positive Ricci curvature.

Since every compact Hermitian symmetric space carries semi-positive holomorphic bisectional curvature and is Kähler-Einstein up to normalizing constants, it is a natural question to ask if such properties of the curvature tensor in fact characterize compact Hermitian symmetric spaces. In order to exploit the semipositivity of bisectional curvatures we state two variational inequalities associated to maxima of holomorphic sectional curvatures and zeros of bisectional curvatures. Let (X,h) be an Kahler manifold of semipositive bisectional curvatures. Fix $x \in X$ and let $\alpha \in T_x^{1,0}(X)$ $\|\alpha\| = 1$ be such that $R_{\alpha\bar{\alpha}\alpha\bar{\alpha}}$ attains the maximum of holomorphic sectional curvatures at x. Then we have (I) For any $\xi \in T_x^{1,0}(X)$ orthogonal to x

$$0 \leq R_{\alpha\bar{\alpha}\xi\bar{\xi}} \leq \frac{1}{2} R_{\alpha\bar{\alpha}\alpha\bar{\alpha}}.$$

Suppose $\beta,\gamma \in T_x^{1,0}(X)$ are such that $R_{\beta\bar{\beta}\gamma\bar{\gamma}} = 0$. Let $\xi \in T_x^{1,0}(X)$ be arbitrary. Then

$$|R_{\beta\bar{\gamma}\xi\bar{\xi}}|^2 + |R_{\beta\bar{\xi}\gamma\bar{\xi}}|^2 \leq R_{\beta\bar{\beta}\xi\bar{\xi}} R_{\gamma\bar{\gamma}\xi\bar{\xi}}.$$

Both (I) and (II) are obtained from computing second variations. For example, letting $F(t) = 1/(1+t^2)^2 \, R(\alpha+t\xi,\ldots,\overline{\alpha+t\xi})$, (I) follows immediately from $F''(t) \leq 0$. For the proof of (II) see Mok-Zhong [16].

(3.2) <u>The formula of Berger--a maximum principle</u>.

The first contribution towards characterizing Hermitian symmetric spaces was Berger's result ([1], 1965) characterizing the complex projective space (equipped with the Fubini-Study metric) by the properties that it is Kähler-Einstein and carries positive Riemannian sectional curvature. He did this by applying the maximum principle to $\alpha \in T^{1,0}(X)$, $\|\alpha\| = 1$, attaining the global maximum of holomorphic sectional curvature, where X is a compact Kähler manifold having both properties. Then he computed

$$\frac{1}{2}\Delta R_{\alpha\bar{\alpha}\alpha\bar{\alpha}} = \Sigma_{k,\ell}|R_{\alpha\bar{k}\alpha\bar{\ell}}| + \rho R_{\alpha\bar{\alpha}\alpha\bar{\alpha}} - 2\Sigma_{k,\ell}|R_{\alpha\bar{\alpha}k\bar{\ell}}|^2$$

where we used an orthonormal basis $\{e_i\}$ of $T_x^{1,0}(x)$, $\alpha \in T_x^{1,0}$, and ρ is the Einstein constant of X. Let g(y) be the smooth function on a neighborhood of x defined by $g_\alpha(y) = R(\alpha(y),\ldots,\overline{\alpha(y)})$ with $\alpha(y)$ obtained from $\alpha(x) = \alpha$ by parallel transport along geodesics emanating from x. Then, we have $\Delta g_\alpha(x) = \Delta R_{\alpha\bar{\alpha}\alpha\bar{\alpha}}$, yielding

$\Delta R_{\alpha\bar{\alpha}\alpha\bar{\alpha}} \leq 0$ by the global maximality of $R_{\alpha\bar{\alpha}\alpha\bar{\alpha}}$. Let now $\{e_i\}$ be an orthonormal basis of $T_x^{1,0}(X)$ consisting of eigenvectors of the Hermitian bilinear form $H(\xi,\xi) = R_{\alpha\bar{\alpha}\xi\bar{\xi}}$. Write $\alpha = e_1$. The preceding formula of Berger simplifies to

$$\frac{1}{2}\Delta R_{1\bar{1}1\bar{1}} = \Sigma_{\substack{k\neq 1 \\ \ell\neq 1}} \text{ or } |R_{1\bar{k},\bar{\ell}}|^2 + \Sigma_{k>1} R_{1\bar{1}k\bar{k}}(R_{1\bar{1}1\bar{1}} - 2R_{1\bar{1}k\bar{k}}),$$

where we replaced ρ by $\Sigma_k R_{1\bar{1}k\bar{k}}$. From this and the inequality (I) of (3.1) we have the opposite inequality $\Delta R_{1\bar{1}1\bar{1}} \geq 0$. Thus, $\Delta R_{1\bar{1}1\bar{1}} = 0$ and $R_{1\bar{1}k\bar{k}} = \frac{1}{2}R_{1\bar{1}1\bar{1}}$ for all $k > 1$. Let now $(W_{i\bar{j}k\bar{\ell}})$ be the Kählerian Bochner-Weyl tensor obtained from $(R_{i\bar{j}k\bar{\ell}})$ simply from $W_{\xi\bar{\xi}\xi\bar{\xi}} = R_{\xi\bar{\xi}\xi\bar{\xi}} - \rho c_n$ (for Kähler-Einstein metrics), where c_n is a constant depending only on $n = \dim_{\mathbb{C}} X$ and chosen such that the average of $W_{\xi\bar{\xi}\xi\bar{\xi}}$ over the unit sphere of $T_y^{1,0}(X)$ is zero for all $y \in X$. W is defined in such a way that X is $\mathbb{CP}^n$ (with the Fubini-Study metric) if and only if W vanishes identically on X. Now the equality $R_{1\bar{1}k\bar{k}} = \frac{1}{2}R_{1\bar{1}1\bar{1}}$ for $k > 1$ can be converted to $W_{1\bar{1}1\bar{1}} = 0$. But $W_{1\bar{1}1\bar{1}}$ is also the global maximum of $W_{\xi\bar{\xi}\xi\bar{\xi}}$, $\|\xi\| = 1$. Thus $W \equiv 0$ and X is projective.

If we assume only that X carries semipositive bisectional curvature we obtain the partial information $R_{1\bar{1}k\bar{k}} = \frac{1}{2} R_{1\bar{1}1\bar{1}}$ or 0 for $k > 1$. While the simplest curvature of characterization of Riemannian locally symmetric spaces by the vanishing of ∇R already involves differentiation, so that the maximum principle of Berger on the unit sphere bundle $S^{1,0}(X)$ of $T^{1,0}(X)$ does not apply directly, we were recently able to derive from it sufficient pointwise information to obtain a global proof of

THEOREM (Mok-Zhong [15][16]): Let X be a compact Kähler-Einstein manifold carrying semipositive holomorphic bisectional curvature. Then X is biholomorphically isometric to a Hermitian symmetric space.

It is tempting to prove our theorem by some form of integral formula in which $\|\nabla R\|^2$ appears as the gradient term. This was in fact done for the special case when X carries semipositive Riemannian sectional curvature long ago by Gray [5]. There he used an integral formula on the unit sphere bundle. However, this formula apparently fails for the situation of semipositive bisectional curvature and the problem was left open (cf. Siu [21]).

The proof of our theorem relies on the maximum principle applied to higher order derivatives and some global techniques of foliation and isometric decomposition of the manifold, which we proceed to prove separately.

(3.3) The maximum principle on higher order derivatives.

The pointwise information obtained for the proof of our theorem can be formulated as follows. Let $S^{1,0}(X)$ be the unit sphere bundle of $T^{1,0}(X)$. We define $\mathfrak{M} \subset S^{1,0}(X)$ to be the set of all $\alpha \in S^{1,0}(X)$ attaining the global maximum of holomorphic sectional curvatures and write $\mathfrak{M}_x = \mathfrak{M} \cap T_x^{1,0}(X)$. Then we have

PROPOSITION:

(1) $\mathfrak{M}_x \neq \emptyset$ for any $x \in X$,

(2) $\nabla_\alpha R = 0$ for a generic $\alpha \in \mathfrak{M}$

We explain first the proof of (1). Let $f\colon X \to \mathbb{R}$ be defined by $f(x) = \sup_{\xi \in S_x^{1,0}(X)} R_{\xi\bar{\xi}\xi\bar{\xi}}$. Then f is continuous and (1) is equivalent to asserting that f is constant. Recall that for $\alpha \in \mathfrak{M}$ we had $0 \geq \Delta R_{\alpha\bar{\alpha}\alpha\bar{\alpha}} \geq \Sigma_{k\geq 1} R_{\alpha\bar{\alpha}k\bar{k}}(R_{\alpha\bar{\alpha}\alpha\bar{\alpha}} - 2R_{\alpha\bar{\alpha}k\bar{k}}) \geq 0$ (cf. (3.2)). The first inequality comes from $\alpha \in \mathfrak{M}$ but the second inequality was obtained from the inequality (I) of (3.1), which only uses the fact that $R_{\alpha\bar{\alpha}\alpha\bar{\alpha}}$ is a pointwise maximum of holomorphic sectional curvatures. Returning to our function $f\colon X \to \mathbb{R}$ we have therefore $\Delta R_{\xi\bar{\xi}\xi\bar{\xi}} \geq 0$ for any $\xi \in T_x^{1,0}(X)$, x arbitrary such that $R_{\xi\bar{\xi}\xi\bar{\xi}} = f(x)$. To prove $f \equiv$ constant it suffices to show $\Delta f \geq 0$ in the sense of distribution. Let $g_\xi\colon U \to \mathbb{R}$ be defined on a small open neighborhood of x by $g_\xi(y) = R(\xi(y),\ldots,\overline{\xi(y)})$, $\xi(y) \in S_y^{1,0}(X)$ obtained from $\xi = \xi(x)$ by parallel transport along a minimal geodesic. Then $\Delta g_\xi(x) = \Delta R_{\xi\bar{\xi}\xi\bar{\xi}} \geq 0$ while $f(x) = g_\xi(x)$ and $f(y) \geq g_\xi(y)$ for $y \in U$. Thus g_ξ is a barrier (from below) of f at x. This proves $\Delta f \geq 0$ on X and hence $f \equiv$ constant by the maximum principle, proving (1) of the proposition.

AVERAGING OPERATORS OF HIGHER DERIVATIVES: For the proof of (2) we need the maximum principle for higher order derivatives. For smooth tensors T we define an elliptic operator $S^{(2k)}T$ of order $2k$ by taking a suitably normalized average of $2k$-th order radial derivatives. Thus

$$S^{(2k)}R_{\alpha\bar{\alpha}\alpha\bar{\alpha}}(x) = c_{2k}\int_{\eta\in S_x(X)} \nabla_\eta^{(2k)} R_{\alpha\bar{\alpha}\alpha\bar{\alpha}}(x)$$

where $S(X)$ denotes the unit sphere bundle of the real tangent bundle $T(X)$ and the integration is performed in terms of the rotationally symmetric unit measure of the sphere $S_x(X) = S(X) \cap T_x(X)$. We choose a constant c_{2k} such that the principal term of $S^{(2k)}T$ with $\Delta^k T$. From Berger's formula we have $\Delta R_{\alpha\bar{\alpha}\alpha\bar{\alpha}} = 0$ for any $\alpha \in \mathfrak{M}$. But Δ is nothing other than $S^{(2)}$, and we conclude thus that $R_{\alpha\bar{\alpha}\alpha\bar{\alpha},\eta\eta} = 0$. By the maximality of $R_{\alpha\bar{\alpha}\alpha\bar{\alpha}}$ we conclude that $R_{\alpha\bar{\alpha}\alpha\bar{\alpha},\eta\eta\eta} = 0$ and that $R_{\alpha\bar{\alpha}\alpha\bar{\alpha},\eta\eta\eta\eta} \leq 0$ so that $S^{(4)}R_{\alpha\bar{\alpha}\alpha\bar{\alpha}} \leq 0$. Suppose we have an analogue of Berger's computation for higher derivatives to conclude $S^{(4)}R_{\alpha\bar{\alpha}\alpha\bar{\alpha}} \geq 0$ we will obtain from $S^{(4)}R_{\alpha\bar{\alpha}\alpha\bar{\alpha}} = 0$ the vanishing of accompanying quantities, which turn out to be first derivatives of the curvature tensor.

Let $\{e_i\}$ be an orthonormal basis of $T_x^{1,0}(X)$ adapted to $\alpha \in \mathfrak{M}$ consisting of eigenvectors of the Hermitian form $H_\alpha(\xi,\zeta) = R_{\alpha\bar{\alpha}\xi\bar{\zeta}}$. We have an orthogonal decomposition $T_x^{1,0}(X) = \mathbb{C}\alpha \oplus \mathcal{H} \oplus \mathcal{N}$ into eigenspaces of H_α with eigenvalues $R_{\alpha\bar{\alpha}\alpha\bar{\alpha}}$, $1/2\, R_{\alpha\bar{\alpha}\alpha\bar{\alpha}}$ and 0 respectively. Write $\alpha = e_1$, $\mathcal{H} = \Sigma_{p\in H}\,\mathbb{C}e_p$ and $\mathcal{N} = \Sigma_{q\in N}\,\mathbb{C}e_q$. To give a complete proof of (2) of our proposition we need to show $\nabla_1 R_{i\bar{j}k\bar{\ell}} = 0$, $1 \leq i,j,k,\ell \leq n = \dim_{\mathbb{C}} X$. Instead of doing this we will simply illustrate the line of thought by the computation $\nabla_1 R_{i\bar{1}k\bar{\ell}} = 0$ for $1 \leq i,k,\ell \leq n$ which already contains the main ideas of the complete proof. By the Bianchi identity we have $\nabla_1 R_{i\bar{1}k\bar{\ell}} = \nabla_i R_{1\bar{1}k\bar{\ell}}$ so that it suffices to show that $\nabla_\eta R_{1\bar{1}k\bar{\ell}} = 0$ for $\eta \in T_x(X)$ a real tangent vector and $1 \leq k,\ell \leq n$.

VARIATION OF THE CURVATURE TENSOR ALONG GEODESICS: Using our special basis it is easy to deduce $R_{1\bar{1}1\bar{1},\eta} = R_{1\bar{1}1\bar{p},\eta} = R_{1\bar{1}1\bar{q},\eta} = R_{1\bar{1}p\bar{p},\eta} = R_{1\bar{1}q\bar{q},\eta} = 0$ for $p \in H$ and $q \in N$. The equations $R_{1\bar{1}1\bar{1},\eta} = R_{1\bar{1}q\bar{q},\eta} = 0$ come simply from the maximality of $R_{1\bar{1}1\bar{1}}$ and the minimality of $R_{1\bar{1}q\bar{q}} = 0$. Consider a geodesic $\gamma(t), -\delta < t < \delta$ with $\gamma(0) = x$ such that $\dot{\gamma}(0) = \eta$. We write $R_{i\bar{j}k\bar{\ell}}(t)$ for $R(e_i(t), \overline{e_j(t)}; e_k(t), \overline{e_\ell(t)})$, the vectors inside the bracket at $\gamma(t)$ being obtained by parallel transport along γ. Consider now the function for $0 \leq t \leq \delta$ defined by

$$F_\sigma(t,\xi) = \frac{1}{(1+t^{2\sigma})^2} R(\alpha+t^\sigma\xi, \ldots, \overline{\alpha+t^\sigma\xi})(t)$$

where ξ is orthogonal to α and $\sigma \geq 0$. By the global maximality of $R_{1\bar{1}1\bar{1}}$ we see that

$$(*) \qquad F_\sigma(t,\xi) \leq R_{1\bar{1}1\bar{1}}(0)(1+t^{2\sigma})^2.$$

Take $\sigma = 1$ and $\xi = e_p$, $p \in H$. Recall that $\nabla^i_\eta R_{1\bar{1}1\bar{1}}(0) = 0$ for $1 \leq i \leq 3$ so that $R_{1\bar{1}1\bar{1}}(t) = R_{1\bar{1}1\bar{1}}(0) + 0(t^4)$. Expanding both sides in Taylor series and comparing coefficients of t^2 we deduce

$$4R_{1\bar{1}p\bar{p}}(0) + 4\mathrm{Re}\, R_{1\bar{1}1\bar{p},\eta}(0) \leq 2R_{1\bar{1}1\bar{1}} = 0.$$

Since we may assume without loss of generality that $R_{1\bar{1}1\bar{p},\eta}$ is real and nonnegative, the preceding inequality yields $R_{1\bar{1}1\bar{p},\eta}(0) = 0$. Similarly comparing Taylor coefficients t^3 yields $R_{1\bar{1}p\bar{p},\eta}(0) = 0$. Furthermore, taking instead $\xi = e_q$, $q \in N$ and $\sigma = 1.5$, say, one can deduce $R_{1\bar{1}1\bar{q},\eta}(0) = 0$.

The discussion in the preceding paragraph shows that the only difficulty in proving $\nabla_\eta R_{1\bar{1}j\bar{k}} = 0$ consists in proving $\nabla_\eta R_{1\bar{1}p\bar{q}} = 0$ for $p \in H$ and $q \in N$. For this we need the following lemma on zero order information of $R_{i\bar{j}k\bar{\ell}}$.

LEMMA:

(i) $R_{1\bar{1}i\bar{j}} = 0$ unless $i = j = 1$

(ii) $R_{1\bar{q}k\bar{\ell}} = 0$ for $q \in N$ and $1 \leq k, \ell \leq n$.

REMARKS: That (i) is valid is suggested by Properties (B) and (C) of the curvature tensor of Hermitian symmetric spaces as given in (3.1). In fact they suggest that $\alpha\circ\alpha$ is an eigenvector of the operator $P\colon S^2T_x^{1,0} \times S^2T_x^{1,0} \to \mathbb{C}$ so that $R_{1\bar{\xi}1\bar{\xi}'} = 0$ for $\xi \circ \xi'$ orthogonal to $\alpha\circ\alpha$. Similarly, (i) is suggested by Property (A). Namely $R_{1\bar{1}q\bar{q}} = 0$ suggests that $\alpha \otimes \bar{e}_q$ is an eigenvector (corresponding to the eigenvalue 0) of the Hermitian, positive semidefinite operator $Q\colon T_x^{1,0} \times T_x^{0,1} \to \mathbb{C}$, so that $R_{1\bar{q}k\bar{\ell}} = 0$ for $1 \leq k, \ell \leq n$.

PROOF OF LEMMA: (i) is a consequence of Berger's computation (cf. (3.2))

$$0 \geq \Delta R_{1\bar{1}1\bar{1}} \geq \Sigma_{\substack{k\neq 1 \text{ or} \\ \ell\neq 1}} |R_{1\bar{k}1\bar{\ell}}|^2 + \Sigma_{k>1} R_{1\bar{1}k\bar{k}}(R_{1\bar{1}1\bar{1}} - 2R_{1\bar{1}k\bar{k}}) \geq 0.$$

(ii) follows from the inequality (II) of (3.1)

$$|R_{1\bar{\xi}q\bar{\xi}}|^2 + |R_{1\bar{q}\xi\bar{\xi}}|^2 \leq R_{1\bar{1}\xi\bar{\xi}} R_{\xi\bar{\xi}q\bar{q}}$$

and a computation of $\Delta R_{1\bar{1}q\bar{q}}$ (≥ 0 since $R_{1\bar{1}q\bar{q}} = 0$ is a minimum of bisectional curvatures).

THE VANISHING OF $\nabla_\eta R_{1\bar{1}p\bar{q}}$.

To prove $\nabla_\eta R_{1\bar{1}p\bar{q}} = 0$ for $p \in H$, $q \in N$ we compute $S^{(4)} R_{1\bar{1}1\bar{1}}$. First, by computing commutations we show that $S^{(4)} R_{1\bar{1}1\bar{1}} - \Delta^2 R_{1\bar{1}1\bar{1}} = \text{const. } \Sigma_{q\in N} \Sigma_{k,\ell} |R_{1\bar{q}k\bar{\ell}}|^2$, which vanishes by (ii) of the lemma. On the other hand $\Delta^2 R_{1\bar{1}1\bar{1}} = S^{(2)} \Delta R_{1\bar{1}1\bar{1}}$. Hence,

$$0 \geq S^{(4)} R_{1\bar{1}1\bar{1}} = c_2 \int_\eta \Delta R_{1\bar{1}1\bar{1},\eta\eta}.$$

We will show that $\Delta R_{1\bar{1}1\bar{1},\eta\eta} \geq 0$ for any $\eta \in S_x(X)$ so that in fact

$\Delta R_{1\bar{1}1\bar{1},\eta\eta} = 0$. From Berger's formula we have

$$\frac{1}{2}\Delta R_{1\bar{1}1\bar{1},\eta\eta} = (\Sigma|R_{1\bar{k}1\bar{\ell}}|^2)_{\eta\eta} - 2(\Sigma|R_{1\bar{1}k\bar{\ell}}|^2)_{\eta\eta} + \rho R_{1\bar{1}1\bar{1},\eta\eta}.$$

From preceding discussion, in particular (i) of the lemma we can simplify to obtain

$$\frac{1}{2}\Delta R_{1\bar{1}1\bar{1},\eta\eta} = 2\Sigma|R_{1k1\ell,\eta}|^2 - 4\Sigma_{p\in H} R_{1\bar{1}p\bar{p}} R_{1\bar{1}p\bar{p},\eta\eta} - 4\Sigma_{\substack{p\in H\\ q\in N}} |R_{1\bar{1}p\bar{q},\eta}|^2$$

Since $R_{1\bar{1}p\bar{p}} = \frac{1}{2} R_{1\bar{1}1\bar{1}}$ and $(R_{1\bar{1}1\bar{1}} + \Sigma_{p\in H} R_{1\bar{1}p\bar{p}} + \Sigma_{q\in N} R_{1\bar{1}q\bar{q}}) = \rho$ implies $\Sigma_{p\in H} R_{1\bar{1}p\bar{p},\eta\eta} = \Sigma_{q\in N} R_{1\bar{1}q\bar{q},\eta\eta}$ we obtain

$$(\#) \qquad \frac{1}{8}\Delta R_{1\bar{1}1\bar{1},\eta\eta} \geq \Sigma_{q\in N}\left(\frac{R_{1\bar{1}1\bar{1}}}{2} R_{1\bar{1}q\bar{q},\eta\eta} - \Sigma_{p\in H}|R_{1\bar{1}p\bar{q},\eta}|^2\right).$$

By the Schwarz inequality we have, developing along a geodesic γ with $\gamma(0) = \eta$ as before

$$|R_{1\bar{1}p\bar{q}}(t)|^2 \leq R_{1\bar{1}p\bar{p}}(t) R_{1\bar{1}q\bar{q}}(t).$$

Expanding in Taylor series and comparing coefficients we obtain

$$|R_{1\bar{1}p\bar{q},\eta}|^2 \leq \frac{R_{1\bar{1}1\bar{1}}}{4} R_{1\bar{1}q\bar{q},\eta\eta}$$

so that from (#) we obtain $\Delta R_{1\bar{1}1\bar{1},\eta\bar{\eta}} \geq 0$ if $\mathcal{H}$ is one-dimensional. In general, one can apply the preceding inequality (suitably normalized) to any $\xi \in H$ in place of e_p. In particular taking $\xi = \Sigma\, \overline{R_{1\bar{1}p\bar{q},\eta}}\, e_p$, we have

$$(\Sigma_{p\in H}|R_{1\bar{1}p\bar{q},\eta}|^2)^2 \leq \frac{R_{1\bar{1}1\bar{1}}}{4}\, \Sigma_{p\in H}|R_{1\bar{1}p\bar{q},\eta}|^2 R_{1\bar{1}q\bar{q},\eta\bar{\eta}} \quad \text{i.e.}$$

$$\Sigma_{p\in H}|R_{1\bar{1}p\bar{q},\eta}|^2 \leq \frac{R_{1\bar{1}1\bar{1}}}{4} R_{1\bar{1}q\bar{q},\eta\bar{\eta}}.$$

We have thus

$$\frac{1}{8}\,\Delta R_{1\bar{1}1\bar{1},\eta\bar{\eta}} \geq 2\Sigma_{q\in N}\left(\frac{R_{1\bar{1}1\bar{1}}}{4} R_{1\bar{1}q\bar{q},\eta\bar{\eta}} - \Sigma_{p\in H}|R_{1\bar{1}p\bar{q},\eta}|^2\right) + \Sigma_{p\in H}|R_{1\bar{1}p\bar{q},\eta}|^2 \geq 0.$$

But then $S^{(4)} R_{1\bar{1}1\bar{1}} = 0$, from which we conclude $\Delta R_{1\bar{1}1\bar{1},\eta\bar{\eta}} = R_{1\bar{1}p\bar{q},\eta} = 0$. This completes the proof of $R_{1\bar{1}k\bar{\ell},\eta} = 0$ for $1 \leq k,\ell \leq n$.

(3.4) Foliating by Hermitian symmetric submanifolds and an isometric decomposition.

Let $V_x \subset T_x^{1,0}(X)$ be the $\mathbb{C}$-linear span of $\mathfrak{M}_x$. One can prove using the Coherence Theorem of Oka that there exists a non-empty open subset U of X such that $V|_U = \cup_{x\in U} V_x$ is a smooth $\mathbb{C}$-vector bundle over U. We assert

PROPOSITION: The distribution $x \to \text{Re}\, V_x$ is integrable on U. Moreover, the integral submanifolds Z_x are totally geodesic, complex, and locally symmetric. Moreover, Z_x can be extended to compact Hermitian symmetric submanifolds $\hat{Z}_x$ of M.

PROOF: By the Theorem of Frobenius and the identity $\nabla_W W' - \nabla_{W'} W - [W,W] = 0$ for smooth vector fields W and W', it suffices to show (the stronger fact) that $\nabla_W W'$ takes values in $\text{Re}\, V|_U$ whenever W and W' are local vector fields on (an open subset of) U taking values in $\text{Re}\, V|_U$. One can easily reduce this to showing that for any curve γ on U with tangent vectors in $\text{Re}\, V|_U$, the parallel transport of a maximal vector α (i.e. $\alpha \in \mathfrak{M}$) remains maximal along γ. But this follows immediately from $\nabla_\alpha R = 0$ for a generic $\alpha \in \mathfrak{M}$. What we have proved is in fact stronger than the integrability of

$\mathrm{Re}\, V|_U$. We actually know that the integral submanifolds Z_x are totally geodesic, complex analytic ($\mathrm{Re}\, V|_U$ is clearly closed under the J-operator) and locally symmetric ($\nabla R \equiv 0$ on Z_x since the tangent spaces to Z_x are generated by real parts of maximal vectors). Moreover Z_x carries positive Ricci curvature. To see this, by the local symmetry of Z_x we can locally decompose Z_x into $Z_x' \times Z_x''$ where Z_x' carries positive Ricci curvature and Z_x'' is flat. But if Z_x'' were non-trivial it is easy to see that $\mathfrak{M}_x$ could not generate $T_x^{1,0}(Z_x)$. Obviously the Ricci curvature on Z_x is parallel. Since Z_x is totally geodesic by analytic continuation using exponential maps Z_x can be completed to some immersed complete Kähler manifold $\hat{Z}_x$ with Ricci curvature bounded from below by a positive constant. By the Theorem of Bonnet-Myers it follows that $\hat{Z}_x$ must be compact. To complete our proposition it suffices to show that $\hat{Z}_x$ has no self-intersections for a suitable choice of U ($\hat{Z}_x$ is then obviously a Hermitian symmetric submanifold since $\nabla R \equiv 0$ on $\hat{Z}_x$ by the principle of analytic continuation.) First there are no tangential self-intersections by the total geodesic property of $\hat{Z}_x$. On the other hand, if two pieces of $\hat{Z}_x$ intersected in a non-tangential way at y then we would have $\dim_{\mathbb{C}} V_y > \dim_{\mathbb{C}} V|_U$. But V_y can be translated to $x \in U$ to give a $\mathbb{C}$-vector subspace of $T_x^{1,0}(X)$ generated by $\alpha \in \mathfrak{M}_x$ strictly larger than V_x, yielding a contradiction.

Finally, we indicate how our proposition can be used to obtain a de Rham decomposition of X. Our argument that $\hat{Z}_x$ has no self-intersection actually proves that $\hat{Z}_x \cap \hat{Z}_{x'} = \emptyset$ whenever $x' \notin \hat{Z}_x$. This implies that the normal bundle of $\hat{Z} = \hat{Z}_x$ in X is differentiably trivial. Let $N^{1,0} = T_X^{1,0}|_{\hat{Z}}/T_{\hat{Z}}^{1,0}$. $N^{1,0}$ is the holomorphic normal bundle of $\hat{Z}$ in X. Writing $N_{\mathbb{R}}$ for the real normal bundle, we have $N^{1,0} \oplus N^{0,1} \cong N_{\mathbb{R}} \otimes_{\mathbb{R}} \mathbb{C}$ differentiably $N^{0,1} = \overline{N^{1,0}}$. Since $N_{\mathbb{R}}$ is differentiably trivial we have in particular $c_1(N^{1,0}) + c_1(N^{0,1}) = 0$ and thus $c_1(N^{1,0}) = 0$. Since quotient bundles of a semipositive Hermitian holomorphic vector bundle are in general semipositive, this implies the pointwise vanishing of $c_1(N^{1,0})$, from which we deduce

(*) $$R_{\xi\bar{\xi}\zeta\bar{\zeta}} = 0 \text{ for } \xi \in T_x^{1,0}(\hat{Z}) \text{ and } \zeta \perp \xi.$$

This is apparently a strong indication that $\hat{Z}$ is a direct factor of an isometric decomposition of X. In fact, we can use this to show that $V|_U$ is invariant under parallel transport. This is obvious along each $\hat{Z}$ but to prove this for parallel transports along geodesics transverse to the foliation of U by Z_x the property (*) is needed. By analytic continuation $V = \cup_{x \in X} V_x$ is a differentiable $\mathbb{C}$-vector bundle invariant under parallel transport. The orthogonal distribution $x \mapsto V_x^{\perp}$ gives rise to a foliation of X by $\hat{Z}_x^{\perp}$. By induction on dimensions one can assume that $\hat{Z}_x^{\perp}$ is Hermitian symmetric, yielding immediately an isometric decomposition $X = X_1 \times X_2$ with X_1 isometric to any $\hat{Z}_x$ and X_2 isometric to any $\hat{Z}_x^{\perp}$.

§4. THE RIGIDITY PHENOMENON OF KÄHLER METRICS OF SEMIPOSITIVE BISECTIONAL CURVATURE ON IRREDUCIBLE COMPACT HERMITIAN SYMMETRIC SPACES OF RANK ≥ 2.

(4.1) Background.

In Mok [13] we studied uniqueness theorems of Hermitian metrics of seminegative curvature on hyperbolic Hermitian symmetric spaces of rank ≥ 2 and of finite volume. In particular, if Ω is an irreducible bounded symmetric domain of rank ≥ 2 and Γ is a discrete group of biholomorphisms of Ω without fixed points such that Ω/Γ is compact, then the Kahler-Einstein metric on Ω/Γ is the unique Hermitian metric (up to multiplicative constants) with seminegative curvature. Write $X = \Omega/\Gamma$. Let g be the Kahler-Einstein metric on X and h be any Hermitian metric of seminegative curvature on X. Denoting by ∇ covariant differentiation on (X,g), the uniqueness theorem is equivalent to asserting the vanishing of ∇h. For this purpose we use an integral formula on some complex submanifold $\overline{\mathfrak{M}} \subset \mathbb{P}T(X)$. In this chapter $T(X)$ will always denote the holomorphic tangent bundle. To be precise let $L \to \mathbb{P}T(X)$ be the line bundle on $\mathbb{P}T(X)$ equipped with induced metrics $\hat{g}$ and $\hat{h}$ derived from g and h, then, for every complex k-dimensional submanifold $\mathcal{S}$ of $\mathbb{P}T(X)$, and every smooth (m,m) form ν_m on $\mathcal{S}$ we have

$$\int_{\mathcal{S}} [c_1(L,\hat{g})]^{k-m} \wedge \nu_m = \int_{\mathcal{S}} [c_1(L,\hat{h})]^{k-m} \wedge \nu_m$$

where c_1 denotes the first Chern form in the given Hermitian metric. To make use of the formula we found a complex submanifold $\bar{\mathfrak{M}} \subset \mathbb{P}T^{1,0}(X)$, an $m > 0$ and a semipositive smooth (m,m) form ν_m such that $\int_{\bar{\mathfrak{M}}} [c_1(L,\hat{g})]^{k-m} \wedge \nu_m = 0$. This forces by the pointwise seminegativity of $c_1(L,\hat{h})$ the formula

$$[c_1(L,\hat{h})]^{k-m} \wedge \nu_m \equiv 0.$$

Denote by R^0 and R the curvature tensors of (X,g) and (X,h) respectively. From the preceding equation we were able to conclude the vanishing of a lot of bisectional curvatures $R_{\xi\bar{\xi}\zeta\bar{\zeta}}$ for which $R^0_{\xi\bar{\xi}\zeta\bar{\zeta}} = 0$. Replacing h by $g + h$ we deduced from comparing R^0 and R the vanishing of a lot of terms of the second fundamental form of the diagonal embedding $(X,g+h) \hookrightarrow (X,g) \times (X,h)$. From such pointwise information we were able to prove directly the vanishing of ∇h.

When one proceeds to ask the analogous question on compact Hermitian symmetric spaces, two difficulties arose. First of all, on any compact Hermitian symmetric space (X,g) (of semipositive holomorphic bisectional curvature) there exists a large group of holomorphic automorphisms which in general do not preserve the metric so that one can only ask about uniqueness up to biholomorphisms. Secondly, there always exist other Hermitian metrics of semipositive curvature. To see this let W be a holomorphic vector field on X and g^* be the dual metric of g on $T^*(X)$, the holomorphic cotangent bundle of X. Then $g^* + 2 \operatorname{Re} W \otimes \bar{W}$ defines a Hermitian metric on $T^*(X)$ which carries seminegative curvature since sums of Hermitian metrics of seminegative curvature remain so. Then the Hermitian manifold $(X,(g^* + 2 \operatorname{Re} W \otimes \bar{W})^*)$ (the dual metric) carries semipositive curvature. This suggests that the correct formulation for metric rigidity phenomena on compact Hermitian symmetric manifolds should be in terms of Kahler metrics. In fact, we proved

THEOREM (Mok [14]): Let (X,g) be an irreducible compact Hermitian symmetric space of rank ≥ 2. Suppose h is a Kähler metric of semipositive holomorphic bisectional curvature. Then (X,h) is also Hermitian symmetric.

While the proof of this theorem is motivated by that of the dual problem for seminegative curvature (Mok [13]), it is global in nature in that an analogous integral formula yields only partial pointwise information which must be "integrated" using the Kähler property of the metric h.

(4.2) An integral formula.

Let (X,g) be an irreducible compact Hermitian symmetric space and let h be a Kähler metric of semipositive holomorphic bisectional curvature. On the holomorphic cotangent bundle $T^*(X)$, the Hermitian metrics g^* and h^* dual to g, h resp. carries seminegative curvature. Let $\Lambda \to \mathbb{P}T^*(X)$ be the projectivisation of $T^*(X)$ and let $\hat{g}^*$, $\hat{h}^*$ be the induced Hermitian metric on Λ. Then, both $(\Lambda,\hat{g}^*)$ and $(\Lambda,\hat{h}^*)$ are pointwise seminegative. Let $\mathfrak{M} \subset T(X)$ be the subset of all unit tangent vectors of type $(1,0)$ realizing the maximum of holomorphic sectional curvature. The contravariant metric tensor $(g^{i\bar{j}})$ defines a bundle mapping from $T(X)$ to $\overline{T^*(X)}$. Denote by Φ the composition of this contraction and conjugation on $\overline{T^*(X)}$, i.e. $\Phi(\Sigma a_i \frac{\partial}{\partial z_i}) = \Sigma_j(\Sigma_i g^{i\bar{j}} a_i) d\bar{z}_j = \Sigma_j(\Sigma_i g^{j\bar{i}} \bar{a}_i) dz_j$ in local holomorphic coordinates. Φ: $T(X) \to T^*(X)$ is antiholomorphic on each fiber. Let $\mathfrak{M}^* = \Phi(\mathfrak{M})$ and denote by $\bar{\mathfrak{M}}^* \subset \mathbb{P}T^*(X)$ the image of $\mathfrak{M}^*$ under the projection map $T^*(X) \to \mathbb{P}T^*(X)$.

By the formula of Berger [1], for any $\alpha \in \mathfrak{M}_x = \mathfrak{M} \cap T_x(X)$ there is an orthogonal decomposition $T_x(X) = \mathbb{C}\alpha \oplus \mathcal{H}_\alpha \oplus \mathcal{N}_\alpha$ of $T_x(X)$ into eigenspaces of the Hermitian form $H_\alpha(\xi,\zeta) = R_{\alpha\bar{\alpha}\xi\bar{\zeta}}$ corresponding to eigenvalues $R_{1\bar{1}1\bar{1}}, \frac{1}{2}R_{1\bar{1}1\bar{1}}$ and 0. We can show that $\dim_{\mathbb{C}}\mathcal{N}_\alpha$, for any $\alpha \in \mathfrak{M}$, is exactly 1 less than the degree of (strong) nondegeneracy of bisectional curvatures of X as computed in Siu [20] and Zhong [25].

We now compute $c_1(\Lambda,\hat{g}^*)$ in terms of the curvature tensor of (X,g). Let $x \in X$ be given. Let $(z_1,\ldots,z_n)$ be a local system of holomorphic coordinates at x and let $\mu \in T_x(X)$ be any unit cotangent vector at x. We can choose $(z_1,\ldots,z_n)$ such that x corresponds to 0, $g_{i\bar{j}}(0) = \delta_{ij}$, $dg_{i\bar{j}}(0) = 0$ and $\mu = \frac{\partial}{\partial z_n}$. Write $[\mu]$ be the corresponding point in $\mathbb{P}T_x(X)$. In terms of inhomogeneous coordinates $(w_1,\ldots,w_{n-1}) = (\xi_1/\xi_n,\ldots,\xi_{n-1}/\xi_n)$ for a cotangent vector $\Sigma\xi_i dz_i$, $\xi_n \neq 0$, $(z_1,\ldots,z_n;w_1,\ldots,w_{n-1})$ is a local system of holomorphic coordinates in a neighborhood of $[\mu]$. The line bundle Λ is holomorphically trivial on the coordinate open set considered and a holomorphic basis can be given by $e(z_1,\ldots,z_n;w_1,\ldots,w_{n-1})$ $= dz_n + \Sigma_{i<n} w_i dz_i \in T^*_z(X) = \Lambda_P(\mathbb{P}T^*(X))$ for $P = [dz_n + \Sigma_{1\le i\le n-1} w_i dz_i]$ $\in \mathbb{P}T_z(X)$. The metric $\hat{g}^*$ is given by

$$\|e\|^2 \overset{\text{def}}{=} \hat{g}^*(e,e) = \Sigma_{i<n}\, g^{i\bar{j}}(z)w_i\bar{w}_j + 2\,\mathrm{Re}\,\Sigma_{i<n} g^{i\bar{n}}(z)w_i + g^{n\bar{n}}(z).$$

For the specific cotangent vector $\mu = \frac{\partial}{\partial z_n}$ at x, μ corresponds to $(0;0)$. Moreover, $\|e([\mu])\|^2 = g^{n\bar{n}}(0) = 1$ and $d_z\|e\|^2[\mu] = d_w\|e\|^2[\mu] = 0$ as can be checked directly using $d_z g_{i\bar{j}}(0) = 0$. The first Chern form $c_1(\Lambda,\hat{g}^*)$ at $[\mu]$ is given by

$$c_1(\Lambda,\hat{g}^*)[\mu] = -\frac{\sqrt{-1}}{2\pi}\,\partial\bar{\partial}\log\|e\|^2[\mu] = -\frac{\sqrt{-1}}{2\pi}\|e\|^2(0;0)$$

$$= -\frac{\sqrt{-1}}{2\pi}\Big(\Sigma_{i<n}\, dw_i \wedge d\bar{w}_i + \Sigma_{i,j}\frac{\partial^2 g^{n\bar{n}}}{\partial z_i\partial\bar{z}_j}(0)\,dz_i \wedge d\bar{z}_j\Big).$$

Since $(z_1,\ldots,z_n)$ is complex geodesic coordinates at x we have

$\frac{\partial^2 g^{n\bar{n}}}{\partial z_i\partial\bar{z}_j}(0) = -R^0_{n\bar{n}i\bar{j}}(0)$ so that $c_1(\Lambda,\hat{g}^*)[\mu]$ is a seminegative $(1,1)$ form at $[\mu]$. (As is apparent from the formula the first Chern form $c_1(L,\hat{g})$ for the induced Hermitian line bundle $L \to \mathbb{P}T(X)$ coming from the holomorphic tangent bundle is <u>not</u> semipositive.)

Recall $\Phi: T(X) \to T^*(X)$ was defined by lifting indices and conjugation. Let now $\alpha \in \mathfrak{M}_x$ and let $s+1$ be the degree of non-degeneracy of bisectional curvatures of X. From the preceding lemma $c_1(\Lambda,\hat{g}^*)[\Phi(x)]$ has exactly s zero eigenvalues. Let ω be the Kahler form of (X,g). Then, we prove in Mok [14]

PROPOSITION 1: $\bar{\mathfrak{M}}^* \subset \mathbb{P}T^*(X)$ is a $2n-s^{-1}$ dimensional complex-analytic submanifold such that for any $[\alpha^*] \in \bar{\mathfrak{M}}^*$, the zero-eigenvectors of $c_1(\Lambda,\hat{g}^*)[\alpha^*]$ are tangent to $\bar{\mathfrak{M}}^*$.

Given the proposition, we have immediately

$$\int_{\bar{\mathfrak{M}}^*}[c_1(\Lambda,\hat{g}^*)]^{2n-2s} \wedge (\pi^*\omega)^{s-1} = 0, \quad \pi: \mathbb{P}T^*(X) \to X \text{ the canonical projection.}$$

Since the integral is a topological invariant we obtain immediately the integral formula (and its consequence)

PROPOSITION 2: Let χ be a Hermitian metric on Λ. We have

$$\int_{\bar{\mathfrak{M}}} [c_1(\Lambda,\chi)]^{2n-2s} \wedge (\pi^*\omega)^{s-1} = 0$$

and hence the vanishing of the integrand everywhere on $\bar{\mathfrak{M}}^*$ whenever (Λ,χ) carries seminegative curvature.

We will now illustrate Proposition 1 (from which one derives Proposition 2) by the special case of hyperquadrics $Q_n = SO(n+2)/SO(n)\times SO(2)$ using explicit curvature computations. (In [14] the proof of Proposition 1 uses some general theory of Hermitian symmetric spaces in place of explicit computations.) We refer to Wolf [24] for general reference on Hermitian symmetric spaces.

The hyperquadric Q_n can be defined as a hypersurface in $\mathbb{P}_{n+1}$ given by $Q_n = \{[z_0,\ldots,z_{n+1}] \in \mathbb{P}_{n+1}: \Sigma z_i^2 = 0\}$. Equipped with the metric g induced by the standard Fubini-Study metric on $\mathbb{P}_{n+1}$ (coming from $\sqrt{-1}\partial\bar\partial \log(\Sigma|z_i|^2)$), (Q_n,g) is a compact Hermitian symmetric space (Q_2,g) is isometrically ($\mathbb{P}_1$, Fubini-Study) $\times$ ($\mathbb{P}_1$, Fubini-Study) while for $n \geq 3$ (Q_n,g) is irreducible. Q_n is a complex-analytic compactification of $\mathbb{C}^n$. One can choose a compactification $\mathbb{C}^n \hookrightarrow Q_n$ such that the translations $z \to z + a$ on $\mathbb{C}^n$ extends to complex-analytic automorphisms of Q_n and such that the metric g restricted to $\mathbb{C}^n$ is given by

$$g = 2 \operatorname{Re} \Sigma \frac{\partial^2 \varphi}{\partial z_i \partial \bar{z}_j} dz_i \wedge d\bar{z}_j, \text{ with}$$

$$\varphi = \log(1 + 2\Sigma|z_i|^2 + |\Sigma z_i^2|^2).$$

The Euclidean coordinate system is a complex geodesic coordinate system for (Q_n,g) at the origin. Then, by computing fourth order derivatives of φ we get

$$R^0_{i\bar{j}k\bar{\ell}} = 4(\delta_{ij}\delta_{k\ell} + \delta_{i\ell}\delta_{kj} - \delta_{ik}\delta_{j\ell})$$

for the curvature tensor $(R^0_{i\bar{j}k\bar{\ell}})$ of (Q_n,g). Then for $\xi = \Sigma\xi_i\frac{\partial}{\partial z_i}$, $\zeta = \Sigma\zeta_i\frac{\partial}{\partial z_i}$ at 0,

$$\frac{1}{4}R^0_{\xi\bar{\xi}\zeta\bar{\zeta}} = \|\xi\|^2\|\zeta\|^2 + |\Sigma\xi_i\bar{\zeta}_i|^2 - |\Sigma\xi_i\zeta_i|^2.$$

By the Schwarz inequality we get $R^0_{\xi\bar{\xi}\zeta\bar{\zeta}} \geq 0$ and that equalities hold if and only if $\xi_i = \bar{\zeta}_i$ and $\Sigma\xi_i^2 = 0$. Moreover, for $\|\xi\| = 1$, $R^0_{\xi\bar{\xi}\xi\bar{\xi}} = 2 - |\Sigma\xi_i^2|^2$ is maximum if and only if $\Sigma\xi_i^2 = 0$. Thus $\bar{\mathfrak{M}}_0$ is nothing other than the complex hyperquadric $Q_{n-2} = \{[\xi_1,\ldots,\xi_n] \in \mathbb{P}_{n-1}: \Sigma\xi_i^2 = 0\}$ and for every $[\alpha] \in \bar{\mathfrak{M}}_0$, the zero eigenspace of $c_1(\Lambda,\hat{g}^*)$ is precisely one-dimensional. Identifying $\mathbb{P}T^*(\mathbb{C}^n)$ with $\mathbb{C}^n \times \mathbb{P}_{n-1}$ in the natural manner, the zero eigenspace of $[\alpha]$ is spanned by $(\bar{\alpha}_1,\ldots,\bar{\alpha}_n;0)$. To prove Proposition 1 it suffices to show $\bar{\mathfrak{M}}^*|_{\mathbb{C}^n} = \mathbb{C}^n \times Q_{n-2}$ under this identification. Since the translation $z \to z + a$ on $\mathbb{C}^n$ extends to complex-analytic automorphisms on Q_n it suffices to show that $\bar{\mathfrak{M}}^*$ is invariant under $\mathrm{Aut}(Q_n)$, the group of complex-analytic automorphisms of Q_n. We now prove this in the general setting of irreducible Hermitian symmetric spaces X.

Write $X = G/K$ as in (3.1) and let $g = k + p$ be the Cartan decomposition of the Lie algebra g of G. Recall that $p_{\mathbb{C}}=p \otimes_{\mathbb{R}} \mathbb{C} = p^+ + p^-$ is the orthogonal decomposition of $p_{\mathbb{C}}$ into eigenspaces of the J-operator and p^+ can be identifies as $T_0(X)$. The complex-analytic group $G_{\mathbb{C}}$ corresponding to $g_{\mathbb{C}} = g \otimes_{\mathbb{R}} \mathbb{C}$ acts holomorphically on X and we have $G/K = G_{\mathbb{C}}/P$ where P is the parabolic complex analytic subgroup of $G_{\mathbb{C}}$ corresponding to the complex Lie subalgebra $k_{\mathbb{C}} + p^-$ where $k_{\mathbb{C}} = k \otimes_{\mathbb{R}} \mathbb{C}$ and P is the identity component of the complex-analytic isotropy subgroup of X at 0. To prove $\bar{\mathfrak{M}}^*$ is invariant under $\mathrm{Aut}(X)$ by the complex analyticity of $\bar{\mathfrak{M}}_0$ and the homogeneity of (X,g) it suffices to observe that p^- acts trivially on p^+ under the adjoint action modulo $k_{\mathbb{C}} + p^-$. (The group $K_{\mathbb{C}} = \exp(k_{\mathbb{C}})$ leaves $\bar{\mathfrak{M}}_0$ invariant.) This follows from $[p^-,p^+] \subset [p_{\mathbb{C}},p_{\mathbb{C}}] \subset k_{\mathbb{C}}$. (The involution σ on g has k and p as eigenspaces corresponding to eigenvalues 1 and -1 so that $\sigma[p_1,p_2] = [\sigma(p_1),\sigma(p_2)] = [-p_1,-p_2] = [p_1,p_2]$ for $p_1,p_2 \in p$, giving $[p,p] \subset k$) and completes the proofs of Propositions 1 and 2 for hyperquadrics Q_n, $n \geq 3$.

(4.3) <u>Pointwise information from the integral formula</u>.

We continue to deal with the special case of Q_n, $n \geq 3$. The degree of non-degeneracy of bisectional curvatures of Q_n is 2 for all n and we have

$$(*) \qquad c_1(\Lambda,\chi)^{2n-2} \equiv 0$$

for χ an arbitrary Hermitian metric of seminegative curvature on Q_n, ω the Kähler form of (Q_n, g) and $\pi: \mathbb{P}T^*(X) \to X$ the canonical projection. Recall that sums of Hermitian metrics of seminegative curvature remain so. We can then take $\chi = \hat{g}^* + \hat{h}^*$ for h a Kähler metric of semipositive holomorphic bisectional curvature on Q_n. Denote by $\overset{\circ}{\Theta}$, Θ and Θ' the curvature forms of g^*, h^* and $g^* + h^*$ on $T^*(Q_n)$ resp. Then, applying (*) to $\chi = \hat{g}^* + \hat{h}^*$ yields immediate $\Theta'_{\zeta^*\bar{\zeta}^*\alpha\bar{\alpha}} = 0$ whenever $\alpha \in \mathfrak{M}$, $\overset{\circ}{\Theta}_{\zeta^*\bar{\zeta}^*\alpha\bar{\alpha}} = 0$. (Then $\zeta^* \in \mathfrak{M}^*$ automatically). We have $\Theta'_{\zeta^*\bar{\zeta}^*\alpha\bar{\alpha}} = \overset{\circ}{\Theta}_{\zeta^*\bar{\zeta}^*\alpha\bar{\alpha}} + \Theta_{\zeta^*\bar{\zeta}^*\alpha\bar{\alpha}} + S_{\zeta^*\bar{\zeta}^*\alpha\bar{\alpha}}$, where S is a nonpositive quantity related to the second fundamental form of the diagonal embedding $(T^*(Q_n), g^* + h^*)$ $(T^*(Q_n), g^*) \times (T^*(Q_n), h^*)$. From this one obtains $\overset{\circ}{\nabla}_\alpha h^{\zeta^*\bar{\tau}^*}$ for $\alpha \in \mathfrak{M}_x$ $\overset{\circ}{\Theta}_{\zeta^*\bar{\zeta}^*\alpha\bar{\alpha}} = 0$ and $\tau^* \in T^*_x(Q_n)$, $\overset{\circ}{\nabla}$ denoting covariant differentiation of (Q_n, g).

To prove (Q_n, h) is Hermitian symmetric by Mok-Zhong [16], it suffices to show that it is Kahler-Einstein. Motivated by Properties (A)-(D) of curvature tensors of irreducible Hermitian symmetric spaces given in (3.1) we prove sequentially for (α, ζ^*) such that $\Theta_{\zeta^*\bar{\zeta}^*\alpha\bar{\alpha}} = 0$

(I) $\Theta_{\zeta^*\bar{\tau}^*\alpha\bar{\sigma}} = 0$ whenever $\sigma \in T_x(Q_n)$, $\tau \in T^*_x(Q_n)$.

(II) $R_{\alpha\bar{\beta}\alpha\bar{\gamma}} = 0$ whenever $\beta \perp \alpha$ or $\gamma \perp \alpha$ (R = curvature tensor of (Q_n, h))

(III) $R_{\alpha\bar{\alpha}\alpha\bar{\alpha}} = c(x)\ h(\alpha\circ\alpha, \alpha\circ\alpha)$

(IV) $\mathrm{Ric}_{\alpha\bar{\beta}} = 0$ for $\beta \perp \alpha$ for the Ricci tensor $(\mathrm{Ric}_{i\bar{j}})$ of (Q_n, h).

(V) (Q_n, h) is Kähler-Einstein.

Here and henceforth $\sigma \perp \tau$ always means $h(\sigma, \tau) = 0$. We are going to explain the reasoning with the exception of (II), the proof of which necessitates "integrating" pointwise information using the Kahler condition along "characteristic spheres" and will be pursued in (4.4).

(I) We work at $0 \in \mathbb{C}^n \subset Q^n$ as usual. Define the operator $Q': (T_0(Q_n) \otimes T^*_0(Q_n))^2 \to \mathbb{C}$ by $Q'_n(\xi \otimes \eta^*, \xi' \otimes \eta'^*) = \Theta_{\eta^*\bar{\eta}'^*\xi\bar{\xi}'}$ and extended by Hermitian bilinearity (I) asserts that $Q'_h(\alpha \otimes \zeta^*, W) = 0$ for any $W \in T_0(Q_n) \otimes T^*_0(Q_n)$. From $\Theta_{\zeta^*\bar{\zeta}^*\alpha\bar{\alpha}} = 0$ we have $Q'_h(\alpha \otimes \zeta^*, \alpha \otimes \tau^*) = 0$ for any $\tau^* \in T^*_0(Q_n)$ by seminegativity of

$\Theta_{\tau*\bar{\tau}*\sigma\bar{\sigma}}$. To prove (I) it suffices to show that the $\mathbb{C}$-span of $\{\alpha \otimes \zeta^* \otimes \bar{\alpha}: \alpha \in \mathfrak{M}, \overset{o}{\Theta}_{\zeta*\bar{\zeta}*\alpha\bar{\alpha}} = 0\}$ contains $\alpha \otimes \zeta^* \otimes T_0(Q_n)$. Recall that $\alpha = \Sigma\alpha_i \frac{\partial}{\partial z_i}$ with $\Sigma\alpha_i^2 = 0$ and $R^0_{\alpha\bar{\alpha}\zeta\bar{\zeta}} = 0$ for $\zeta = \Sigma\,\bar{\alpha}_i \frac{\partial}{\partial z_i}$. By lifting ζ and conjugating, we have $\overset{o}{\Theta}_{\zeta*\bar{\zeta}*\alpha\bar{\alpha}} = 0$ for $\zeta^* = \Sigma\,\alpha_i dz_i$. Fix a pair $(\alpha_0,\bar{\zeta}_0^*)$ and expand $\alpha \otimes \zeta^* \otimes \bar{\alpha}$ in Taylor series at the point $\alpha_0 \otimes \zeta_0^* \otimes \bar{\alpha}_0$. One can use holomorphic coordinates such that both α and ζ^* varies holomorphically while $\bar{\alpha}$ varies anti-holomorphically. By extracting Taylor coefficients it is immediate that $\alpha_0 \otimes \zeta_0^* \otimes \bar{\tau} \in \mathbb{C}\{\alpha \otimes \zeta^* \otimes \bar{\alpha}: \alpha \in \mathfrak{M}, \overset{o}{\Theta}_{\zeta*\bar{\zeta}*\alpha\bar{\alpha}} = 0\}$ as long as τ lies in the $\mathbb{C}$-span of α. But since Q_n is irreducible the $\mathbb{C}$-span of $\mathfrak{M}_0$ is $T_0(Q_n)$. This proves (I).

(III) Assume (II). Then, $\alpha\circ\alpha$ is an eigenvector of the Hermitian bilinear form $P_h: S^2T_0(Q_n) \times S^2T_0(Q_n) \to \mathbb{C}$ defined using h as in (3.1). But since $\mathfrak{M}$ is connected $\{\alpha\circ\alpha: \alpha \in \mathfrak{M}\}$ describes a continuous family of eigenvectors of P_h, so that $P_h(\alpha\circ\alpha,\alpha\circ\alpha) = c(x)h(\alpha\circ\alpha,\alpha\circ\alpha)$ with $c(x)$ independent of x.

(IV) To prove (IV) it is first necessary to give a description of the decomposition $T_0(Q_n) = \mathbb{C}\alpha \oplus \mathcal{H}_\alpha \oplus \mathcal{N}_\alpha$. Recall $Q_n = SO(n+2)/SO(n)\times SO(2)$. It is easy to see that $SO(n)\times SO(2)$ acts transitively on $\bar{\mathfrak{M}}$ so that we may take without loss of generality $\alpha = \frac{1}{\sqrt{2}}(\frac{\partial}{\partial z_{n-1}} + \sqrt{-1}\,\frac{\partial}{\partial z_n})$, $\zeta = \frac{1}{\sqrt{2}}(\frac{\partial}{\partial z_{n-1}} - \sqrt{-1}\,\frac{\partial}{\partial z_n})$ and $\mathcal{H}_\alpha = \mathbb{C}\,\frac{\partial}{\partial z_1} +\ldots+ \mathbb{C}\,\frac{\partial}{\partial z_{n-2}}$. Since (Q_n,g) satisfies $g_{i\bar{j}}(0) = \delta_{ij}$, we have $\zeta^* = \frac{1}{\sqrt{2}}(dz_{n-1} + \sqrt{-1}\,dz_n)$. Thus $\zeta^*(\alpha) = \zeta^*(\xi) = 0$ for any $\xi \in \mathcal{H}_{\alpha_0}$. Write $e_q^\# \in T_0(Q_n)$ for the unit vector relative to h obtained by contracting $\bar{\zeta}^*$ using $(h_{i\bar{j}})$. Then $e_q^\# \perp (\mathbb{C}\alpha + \mathcal{H}_\alpha)$ since $h(\eta,e_q^\#) = \zeta^*(\eta)$ for all $\eta \in T_0(Q_n)$. Let $\{e_1\} \cup \{e_p^\#: p^\# \in H^\#\}$ be an orthonormal basis of $\mathbb{C}_\alpha + \mathcal{H}_\alpha$ relative to h. Then $\{e_1\} \cup \{e_p^\#: p^\# \in H^\#\} \cup \{e_q^\#\}$ is an orthonormal basis of $T_0(Q_n)$ relative to h. Using this basis it suffices for (IV) to prove $Ric_{1\bar{p}^\#} = Ric_{1\bar{q}^\#} = 0$. Now from (I) we deduce $Ric_{1\bar{q}^\#} = 0$ readily. Observe further that $Ric_{1\bar{p}} = R_{1\bar{p}1\bar{1}} + R_{1\bar{p}q^\#\bar{q}^\#} + \Sigma_{r^\#\in H^\#} R_{1\bar{p}^\# r^\#\bar{r}^\#}$. In view of (I) and (II) it suffices by polorization to show that $R_{1\bar{p}^\# p^\#\bar{p}^\#} = 0$ for any $p^\# \in H^\#$

(and any choice of $e^{\#}_p$). Now $\bar{\mathfrak{M}}$ can be expanded in a neighborhood of α by

$$[\alpha(\xi)] = [\alpha + \xi + \frac{1}{2}(\Sigma\xi_i^2)\zeta]$$

where in a neighborhood of α, $\bar{\mathfrak{M}}$ is parametrized by $\xi = \Sigma_{i=1}^{n-2} \xi_i \frac{\partial}{\partial z_i}$. By expanding (III) along a curve $[\alpha + te^{\#}_p + O(t^2)\cdot\zeta]$ in $\bar{\mathfrak{M}}$ we obtain

$$R_{1\bar{1}1\bar{1}} + (4 \text{ Re } R_{1\bar{1}1\bar{p}^{\#}})t + (4R_{1\bar{1}p^{\#}\bar{p}^{\#}} + 4 \text{ Re } R_{1\bar{p}^{\#}1\bar{p}^{\#}})t^2$$

$$+ (4 \text{ Re } R_{1\bar{p}^{\#}p^{\#}\bar{p}^{\#}})t^3 + \ldots = R_{1\bar{1}1\bar{1}}(1 + 2t^2 + t^4).$$

Comparing coefficients of t^3 we get Re $R_{1\bar{p}^{\#}p^{\#}\bar{p}^{\#}} = 0$. Replacing $e^{\#}_p$ by $e^{i\theta}e^{\#}_p$ we may assume $R_{1\bar{p}^{\#}p^{\#}\bar{p}^{\#}}$ is real. Thus, $R_{1\bar{p}^{\#}p^{\#}\bar{p}^{\#}} = 0$, hence $\text{Ric}_{1\bar{p}^{\#}} = 0$, proving $\text{Ric}_{\alpha\bar{\beta}} = 0$ whenever $\beta \perp \alpha = e_1$.

(V) By (IV) every $\alpha \in \mathfrak{M}_x$ is an eigenvector of the Ricci form. But since $\mathfrak{M}_x$ is connected and the $\mathbb{C}$-span of $\mathfrak{M}_x$ is the whole space $T_x(Q_n)$ we conclude that the Ricci form is proportional to the Kähler form ν of (Q_n,h). Thus Ric $= f\nu$. But since the Ricci form is closed we obtain from $d(f\nu) = 0$ that f is a constant. This proves (V) that (Q_n,h) is Kahler-Einstein and hence Hermitian symmetric by Mok-Zhong [16], granted (II).

(4.4) <u>Characteristic spheres and the Kahler condition.</u>

Let (X,g) be an irreducible Hermitian symmetric space of rank ≥ 2. We define a characteristic sphere $S \subset X$ to be an embedded Riemann sphere totally geodesic in (X,g) such that for each $x \in S$ the holomorphic tangent space $T_x(S)$ is generated by some $\alpha \in \mathfrak{M}_x$. For Q_n, in terms of standard coordinates used above, at $0 \in \mathbb{C}^n \subset Q_n$, every $\alpha \in \mathfrak{M}$ defines a characteristic sphere $S_\alpha \subset Q_n$ such that $S_\alpha \cap \mathbb{C}^n$ is the complex line $\{\lambda\alpha: \lambda \in \mathbb{C}\}$. We observe that $\bar{\mathfrak{M}}$ does not depend on the specific choice of a Hermitian symmetric background metric g since $\bar{\mathfrak{M}}$ is complex analytic, thus invariant under the action of $\text{Aut}(Q_n)$. Moreover, the S_α are actually totally geodesic in (Q_n,Ψ^*g) for a complex-analytic automorphism Ψ. To prove (II) we first reduce it to the geometric statement.

(*) Let $S \subset Q_n$ be a characteristic sphere. Then S is totally geodesic in (Q_n,h) for any Kähler metric h of semipositive holomorphic bisectional curvature on Q_n.

We are going to deduce (II) from (*). It is immediate from (*) that $R_{\alpha\bar{\alpha}\alpha\bar{\beta}} = 0$ for $\alpha \in T_x(S)$ and $\beta \perp \alpha$. Since for any $\alpha \in \mathfrak{M}_x$ there exists a characteristic sphere passing through x with $T_x(S) = \mathbb{C}\alpha$, by expanding $R_{\alpha\bar{\alpha}\alpha\bar{\beta}} = 0$ in a neighborhood of some $[\alpha_0]$ in $\mathfrak{M}$ we can deduce (II) just as in the proof of (I).

PROOF OF (*): Fix $[\alpha] \in \bar{\mathfrak{M}}_x$ and let S be a characteristic sphere passing through x such that $T_x(S) = \mathbb{C}\alpha$. To prove (*) it suffices to show that for all choices of x and α, and for $\Gamma_{ij,\bar{k}}$ the Riemann-Christoffel symbols of (Q_n,h), we have

(#) $\Gamma_{\alpha\alpha,\bar{\eta}} = 0$ for all $\eta \perp \alpha$, $\eta \in T_x(X)$.

Without loss of generality we may take $x = 0$. Recall the g-orthogonal decomposition $T_0(Q_n) = \mathbb{C}\alpha + \mathcal{H}_\alpha + \mathcal{N}_\alpha$ given in §3. Let $T_0^*(Q_n) = \mathbb{C}\alpha^* + \mathcal{H}_\alpha^* + \mathcal{N}_\alpha^*$ be the g-orthogonal decomposition of $T_0^*(Q_n)$ dual to that of $T_0(Q_n)$ with respect to g. We have $\tau^*(\alpha) = 0$ for $\tau^* \in T_0^*(Q_n)$ if and only if $\tau^* \in \mathcal{H}_\alpha^* + \mathcal{N}_\alpha^*$. On the other hand (#) is equivalent by h-duality to the statement that $\Gamma_{\alpha\alpha}^{\tau^*} = 0$ for all τ^* such that $\tau^*(\alpha) = 0$. Using the basis $\{e_1\} \cup \{e_p\}_{p\in H} \cup \{e_q\}_{q\in N}$, $e_1 = \alpha$, of $T_0(Q_n)$ adapted to α as was done in (3.3), to prove (#) it suffices to show

(#) $\Gamma_{11}^p = \Gamma_{11}^q = 0$ for all $p \in H$ and all $q \in N$.

At 0, recall from (4.3) $\nabla_1^o h^{q\bar{j}} = 0$ for all j and for the covariant differentiation ∇^o of (Q_n,g). But since the Euclidean coordinate system is complex geodesic for (Q_n,g) at 0, we have $h^{q\bar{j}}{}_{,1} \overset{\text{def}}{=} \dfrac{\partial h^{q\bar{j}}}{\partial z_1} = 0$ at the origin. Thus

$$\Gamma_{11}^q = \Sigma_j\, h^{q\bar{j}} h_{1\bar{j},1} = -\Sigma_j\, h^{q\bar{j}}{}_{,1} h_{1\bar{j}} = 0.$$

For the proof of $\Gamma_{11}^p = 0$, $p \in H$, we are going to use the Kähler property of (Q_n,h). First, observe that we have the expansions

$$[\alpha(\xi)] = [e_1 + \Sigma_{p\in H}\, \xi_p e_p + \frac{1}{2}(\Sigma\xi_p^2) e_q]$$

$$[\zeta^*(\xi)] = [e^q + \Sigma_{p\in H}\, \xi_p e^p + \frac{1}{2}(\Sigma\xi_p^2) e^1]$$

for (α,ζ^*) with $[\alpha] \in \bar{\mathfrak{M}}$ and $\overset{0}{\Theta}_{\zeta^*\bar{\zeta}^*\alpha\bar{\alpha}} = 0$, in terms of the g-orthonormal basis of $T_0(Q_n)$ described in (IV). The equation $h^{\zeta^*\bar{j}}{}_{,\alpha} = 0$ implies $\Sigma\, a_{ik} h^{i\bar{j}}{}_{,k} = 0$ whenever $\Sigma\, a_{ik} e_i \otimes e^k$ is in the $\mathbb{C}$-linear span of $\{\alpha \otimes \zeta^*: \alpha \in \mathfrak{M},\ \overset{0}{\Theta}_{\zeta^*\bar{\zeta}^*\alpha\bar{\alpha}} = 0\}$. In particular by expanding $\alpha \otimes \zeta^*$ we obtain $h^{p\bar{j}}{}_{,1} + h^{q\bar{j}}{}_{,p} = 0$. Thus, for $p \in H$,

$$\Gamma^p_{11} = \Sigma_j\, h^{p\bar{j}} h_{1\bar{j},1} = -\Sigma_j\, h^{p\bar{j}}{}_{,1} h_{i\bar{j}} = \Sigma_j\, h^{q\bar{j}}{}_{,p} h_{i\bar{j}} = -\Sigma_j\, h^{q\bar{j}} h_{1\bar{j},p}.$$

By the Kähler condition we obtain from $h_{1\bar{j},p} = h_{p\bar{j},1}$

$$\Gamma^p_{11} = -\Sigma h^{q\bar{j}} h_{p\bar{j},1} = \Sigma\, h^{q\bar{j}}{}_{,1} h_{p\bar{j}} = 0.$$

Thus $\Gamma_{11,\bar{\eta}} = 0$ whenever $\eta \perp \alpha = e_1$. The proof of (*) and hence of our theorem for hyperquadrics is completed.

BIBLIOGRAPHY

[1] Berger, M. Sur les variétés d'Einstein compactes, C.R. IIIe Reunion Math., Expression Latine, Namur (1965), 35-55.

[2] Borel, A. On the curvature tensor of the Hermitian symmetric manifolds, Ann. of Math. 71 (1970), 508-521.

[3] Borel, A., Density properties for central subgroups of semisimple groups without compact components, Ann. of Math. 72 (1960), 179-188.

[4] Calabi, E. and Vesentini, E., On compact locally symmetric Kahler manifolds, Ann. of Math 71 (1960), 472-507.

[5] Gray, A., Compact Kähler manifolds with nonnegative sectional curvature, Invent. Math. 41 (1977), 33-43.

[6] Jost, J. and Yau, S-T., Harmonic mappings and Kähler manifolds, Math. Ann. 262 (1983), 145-166.

[7] Jost, J. and Yau, S.-T., A strong rigidity theorem for a certain class of compact analytic surfaces. Preprint.

[8] Lai, K.-F. and Mok, N., On a vanishing theorem on irreducible quotients of finite volume of polydiscs. Preprint.

[9] Matsushima, Y. and Shimura, G., On the cohomology groups attached to certain vector-valued differential forms on products of the upper half plane, Ann. of Math. 78 (1963), 417-449.

[10] Mok, N., The holomorphic or anti-holomorphic character of harmonic maps into irreducible compact quotients of polydiscs. Preprint.

[11] Mok, N., La rigidité forte des quotients compacts des polydisques en terme des groupes fondamentaux. Preprint.

[12] Mok, N., Metric rigidity theorems on locally symmetric Hermitian spaces. Preprint.

[13] Mok, N., Uniqueness theorems of Hermitian metrics of seminegative curvature on locally symmetric Hermitian spaces of noncompact type, to appear.

[14] Mok, N., Uniqueness theorems of Kähler metrics of semipositive bisectional curvature on compact Hermitian symmetric spaces, to appear.

[15] Mok, N. and Zhong, J.-Q., Variétés compactes kählériennes d'Einstein de contbure bisectionnelle semipositive, to appear in C.R. Acad. Sci., Paris, 1984.

[16] Mok, N. and Zhong, J.-Q., Curvature characterization of compact Hermitian symmetric spaces, to appear in J. Diff. Geom.

[17] Mostow, G.D., Strong rigidity of locally symmetric spaces, Ann. Math. Studies 78, Princeton; Princeton University Press 1973.

[18] Siu, Y.-T., The complex-analyticity of harmonic maps and the strong rigidity of compact Kähler manifolds, Ann. of Math. 112 (1980), 73-111.

[19] Siu, Y.T., Strong rigidity of compact quotients of exceptional bounded symmetric domains, Duke Math. J. 48 (1981), 857-871.

[20] Siu, Y.-T., Complex-analyticity of harmonic maps, vanishing and Lefschetz theorems, J. Diff. Geom. 17 (1982), 55-138.

[21] Siu, Y.-T.,Some recent developments in complex differential geometry. Lecture at the International Congress of Mathematicians, Warsaw, 1983. Preprint.

[22] Siu, Y.-T., Some recent results in complex manifold theory related to vanishing theorems for the semi-positive case, to appear in a special volume on the Arbeitstagung, Bonn, June 1984.

[23] Weil, A., On discrete subgroups of Lie groups: II, Ann. of Math., 75 (1962), 578-602.

[24] Wolf, J.A., Fine structure of Hermitian symmetric spaces, Geometry of Symmetric Spaces, edited by Bobthhy-Weiss, Marcel-Dekker, New York, 1972.

[25] Zhong, J.-Q., The degree of strong nondegeneracy of the bisectional curvature of exceptional bounded symmetric domains, Proc. Intem. Conf. Several Complex Variables. Hangzhou, edited by Kohn-Lu-Remmert-Siu, Birkhauser, Boston 1984.

DEPARTMENT OF MATHEMATICS
COLUMBIA UNIVERSITY
NEW YORK, NEW YORK 10027

Contemporary Mathematics
Volume **49**, 1986

COMPLEX FINSLER METRICS

H. L. Royden[1]

ABSTRACT. A complex Finsler metric on a complex manifold M is a form $F(\xi,x)$ which gives a norm on the tangent vector ξ at x. We define the holomorphic sectional curvature at (ξ,x) to be the supremum of the Gauss curvature of the metric $F(\dot{\varphi},\varphi)$ for all maps φ of a disk into M with $\varphi(0) = x$, $\dot{\varphi}(0) = \xi$. We give formulas for this curvature when F is sufficiently smooth and consider extensions to the non-smooth case.

1. FINSLER AND FINSLER-TYPE METRICS. By a <u>Finsler metric</u> F on a differentiable manifold M we mean a non-negative $F(\xi,x)$ defined on the tangent bundle TM which satisfies

1) $$F(\alpha\xi,x) = |\alpha| F(\xi,x)$$

2) $$F(\xi_1+\xi_2) \leq F(\xi_1,x) + F(\xi_2,x) .$$

In addition we suppose that F satisfies some regularity conditions, or at least that it is Borel measurable. We also assume that $F(\xi,x) > 0$ for $\xi \neq 0$. If we have a form F which satisfies (1) and the regularity conditions, we say that F is a differential metric of <u>Finsler-type</u>. We define the indicatrix (at x) of a Finsler-type metric F to be the set

$$I = I_x(F) = \{\xi \in T_x : F(\xi,x) < 1\} .$$

Thus a Finsler-type metric is a Finsler metric iff its indicatrix is always convex.

A Finsler-type metric F defines a length function for smooth curves by

$$\ell(\gamma) = \int_a^b F(\dot{\varphi}(t),\varphi(t))dt ,$$

where $\varphi: [a,b] \to M$ is a differentiable parametrization of γ. This length function defines a pseudo-metric ρ by

---

1980 Mathematics Subject Classification 32H15, 53B35, 53B40.

[1] This work was supported by the National Science Foundation, grant no. MCS-83-01379.

$$\rho(p,q) = \inf\{\ell(\gamma): \gamma \text{ joins } p \text{ to } q\} .$$

If F is continuous, then ρ will be a metric. It is called the integrated form of F.

To each Finsler-type metric F there is a supporting Finsler metric $\hat{F}$ which is the largest Finsler metric with $\hat{F} \leq F$. The indicatrix of $\hat{F}$ is the convex hull of the indicatrix of F. Observe that if F is upper semi-continuous then F and $\hat{F}$ have the same integrated metric.

For a complex manifold M, we say that F is a complex Finsler-type metric if (1) holds for all complex α. If (2) holds also, we speak of a complex Finsler metric. Every Hermitian metric is a complex Finsler metric. Other examples of complex Finsler metrics are the Carathéodory-Reiffen metric for complex manifolds and the Teichmüller metric on Teichmüller space. The Kobayashi metric is the integrated form of a Finsler-type metric.

2. CURVATURE FOR ONE DIMENSIONAL METRICS. If M is a one dimensional complex manifold (i.e., a Riemann surface), then a Finsler-type metric is necessarily of the form

$$ds = \lambda|d\zeta| , \tag{3}$$

and hence is a Kähler metric. Such metrics are often called conformal metrics.

If such a metric is of class C^2 at a point ζ_0, we define the Gauss curvature at ζ_0 by

$$K(\lambda) = -4\lambda^{-2} \frac{\partial^2}{\partial\zeta\partial\bar{\zeta}} (\log \lambda) \tag{4}$$

If λ is not of class C^2 at ζ_0, we may still define curvature using supporting metrics. A conformal metric $\hat{\lambda}$ is said to be a <u>supporting metric</u> for λ at ζ_0 if $\hat{\lambda}(\zeta) \leq \lambda(\zeta)$ in some neighborhood of ζ_0, and $\hat{\lambda}(\zeta_0) = \lambda(\zeta_0)$. For $\lambda|d\zeta|$, we define the upper curvature $\overline{K}(\lambda)$ by

$$\overline{K}(\lambda) = \inf K(\hat{\lambda}) \tag{5}$$

where $\hat{\lambda}$ is a C^2 supporting metric for λ at ζ_0. Similarly,

$$\underline{K}(\lambda) = \sup K(\hat{\lambda}) \tag{6}$$

where $\hat{\lambda}$ ranges over all C^2 metrics which are supported by λ at ζ_0. If λ is C^2 at ζ_0, then $\overline{K}$ and $\underline{K}$ are both equal to the Gauss curvature of λ at ζ_0. If $\hat{\lambda}$ is an arbitrary supporting metric for λ at ζ_0, then we have

$$\overline{K}(\lambda) \leq \overline{K}(\hat{\lambda}); \quad \underline{K}(\lambda) \leq \underline{K}(\hat{\lambda}) \tag{7}$$

at ζ_0. If $\overline{K} = \underline{K}$ at ζ_0, we call their common value the curvature of λ at ζ_0. If $\overline{K}(\lambda) \leq a$ everywhere, we say λ has curvature less than or equal to a, and similarly for $\underline{K}(\lambda) \geq b$.

3. HOLOMORPHIC SECTIONAL CURVATURE. Let F be a sufficiently smooth Finsler or Finsler-type metric on a complex manifold M, and let φ be a holomorphic map of a disk Δ_a into M. Then

$$\varphi^*f = F(\dot{\varphi}(\zeta), \varphi(\zeta))|d\zeta|$$

defines a conformal metric on Δ_a. We define the holomorphic sectional curvature $K(F)$ at (x,ξ) to be the supremum of the Gauss curvature at $\zeta = 0$ of $\varphi^*(F)$ as φ ranges over all maps of a disk into M with $\varphi(0) = x$ and $\dot{\varphi}(0) = \xi$. If K is not smooth, we define $\overline{K}(F)$ as the supremum of $\overline{K}(\varphi^*F)$ and $\underline{K}(F)$ as the supremum of $\underline{K}(\varphi^*F)$

PROPOSITION 1. Let M be a complex manifold which is complete and hyperbolic for the Kobayashi metric F_k. Then $\underline{K}(F_K) \geq -4$.

PROOF. Let $(\xi,x) \in TM$ with $F_K(\xi,x) = 1$. Since M is complete hyperbolic there is an "extremal disk" for (ξ,x), i.e., a map φ of the unit disk into M with $\varphi(0) = x$, $\dot{\varphi}(0) = \xi$. Thus $\underline{K}(F_K) \geq \underline{K}(\varphi^*F)$. Since the Kobayashi metric decreases under holomorphic maps, $\varphi^*F \leq (1 - |\zeta|^2)^{-1}$. Thus φ^*F is a supporting metric for the metric $(1 - |\zeta|^2)^{-1}$, and so $\underline{K}(\varphi^*F) \geq -4$. ∎

PROPOSITION 2. Let M be a complex manifold which is Carathéodory hyperbolic, and let F_c be the Carathéodory-Reiffen metric. Then $\overline{K}(F_c) \leq -4$.

PROOF. Let $(\xi_0,x_0) \in TM$ with $F_c(\xi_0,x_0) = 1$. Since M is hyperbolic, there is a holomorphic map f of M into the unit disk Δ such that $f(x_0) = 0$ and $f(\xi_0)$ is the unit tangent vector to Δ. Let φ be a holomorphic map of Δ_a into M with $\varphi(0) = x_0$, $\dot{\varphi}(0) = \xi_0$. Since f is distance decreasing for the Carathéodory-Reiffen metric, we have

$$\varphi^*F_c = F_c(\dot{\varphi}(\zeta), \varphi(\zeta)) \geq \frac{|(f \circ \varphi)'|}{1 - |f \circ \varphi|^2} = \hat{\lambda}.$$

Thus the metric $\hat{\lambda}$ supports φ^*F at $\zeta = 0$. Since $K(\hat{\lambda}) = -4$, $\overline{K}(\varphi^*F) \leq -4$, and we see that $\overline{K}(F_c) \leq -4$. ∎

Since $F_K \geq F_c$, we see that at any place (ξ,x) where they are equal F_c is a supporting metric for F_K and hence they both must have holomorphic sectional curvature equal to -4.

4. FORMULAS FOR SMOOTH METRICS. Let F be a smooth metric of Finsler-type. In order to compute expressions for curvature it is convenient to use the form $G(\xi,x) = F^2$. This form is homogeneous of degree 2 in ξ. The holomorphic sectional curvature of F is given by

$$K = \sup - |G(\dot{\varphi}(\zeta), \varphi(\zeta))]_{\zeta\bar{\zeta}}|_{\zeta=0} \tag{8}$$

where φ ranges over all holomorphic maps φ of a small disk into M with $\dot{\varphi}(0) = x$ and $\varphi(0) = \xi/\,\|\xi\|^2$. Henceforth we shall assume that $\|\xi\| = 1$.

If we let subscripts α, $\overline{\beta}$, etc., denote derivatives with respect to ξ^α, $\overline{\xi^\beta}$, etc., then (8) becomes

(9) $$K = -2\inf\,[G_{\alpha\overline{\beta}}\eta^\alpha\overline{\eta^\beta} + 2\mathrm{Re}\, G_{x^\delta\overline{\beta}}\xi^\delta\overline{\eta^\beta} + G_{x^\gamma\overline{x^\delta}}\xi^\gamma\overline{\xi^\delta}]\,,$$

where $\eta^\alpha = \ddot{\varphi}^\alpha$. If $G_{\alpha\overline{\beta}}$ is positive definite, i.e. if the indicatrix of F is strongly pseudoconvex, then the infinimum is attained when

(10) $$\ddot{\varphi}^\alpha = -G^{\alpha\overline{\beta}}\,G_{\overline{\beta}x^\gamma}\xi^\gamma\,,$$

and its value is given by

(11) $$K = -2G_\alpha(G^{\alpha\overline{\beta}}\,G_{\overline{\beta}x^\gamma})_{\overline{x^\zeta}}\,\xi^\gamma\overline{\xi^\delta}\,,$$

where we have taken $\|\xi\| = 1$. If $G_{\alpha\overline{\beta}}$ is indefinite, $K = +\infty$. If $G_{\alpha\overline{\beta}}$ is positive semi-definite, the value of K may be $+\infty$, depending on $G_{x^\zeta\overline{\beta}}$.

If we set

(12) $$\Gamma^\alpha_\gamma = G^{\alpha\overline{\beta}}\,G_{\overline{\beta}x^\gamma}\,,$$

then

(13) $$K = -2G_\alpha(\Gamma^\alpha_\gamma)_{\overline{x^\delta}}\,\xi^\gamma\overline{\xi^\delta}\,.$$

The quantity $\Gamma^\alpha_{\gamma,\varepsilon}$ is not a tensor, but changes coordinates by adding the term

$$\frac{\partial^2 y^\alpha}{\partial x^\gamma\partial x^\varepsilon}\,.$$

Thus it probably deserves to be called the conformal connection of the metric. If G is Hermitian

$$G = g_{\alpha\overline{\beta}}\xi^\alpha\overline{\xi^\beta}\,,$$

then $\Gamma^\alpha_{\gamma,\varepsilon}$ agrees with the Cartan-Chern connection

(14) $$g^{\alpha\overline{\beta}}g_{\gamma\overline{\beta},\varepsilon}\,,$$

and we see that for a Hermitian metric our definition of holomorphic sectional curvature is just the value of the curvature tensor of the form (14) on the vector $\xi/\,\|\xi\|$.

Given a point $(\xi,x) \in TM$, we may always choose local coordinates so that

$$\Gamma^\alpha_{\gamma,\varepsilon} = -\Gamma^\alpha_{\varepsilon,\gamma}$$

at that point. We call these semi-normal coordinates at (x,ξ). The expression

(10) for $\ddot{\varphi}^\alpha$ maximizing the curvature of φ^*F reduces to $\ddot{\varphi}^\alpha = 0$ in semi-normal coordinates. Thus in these coordinates the holomorphic sectional curvature at (ξ,x) is just the Gauss curvature of the metric restricted to the (complex) line in the direction ξ.

For a Kähler metric, the skew part of $\Gamma^\alpha_{\varepsilon,\gamma}$ is zero, and so $\Gamma^\alpha_{\varepsilon,\gamma}$ vanishes at x for semi-normal coordinates at x. In this case we call the coordinates normal, and K is just the Riemann sectional curvature for the section spanned by ξ and $i\xi$.

This is no longer the case for Hermitian metrics, as the following example shows: Define a metric on the unit ball in $\mathbb{C}^n$ by

$$ds^2 = \frac{\|dz\|^2}{(1 - \|z\|^2)^2} .$$

This is a Hermitian metric, and its Riemann sectional curvature is everywhere -4, but the holomorphic sectional curvature at (ξ,z) is given by

$$K = -4(1 - \|z\|^2 + (\xi\cdot z)^2/\|\xi\|^2) .$$

This K varies from -4 for radial ξ to $-4(1 - \|z\|^2)$ for $(\xi\cdot z) = 0$. I am indebted to Charles Stanton for this example.

5. KÄHLER CONDITIONS FOR FINSLER METRICS. There is an analogue of the Kahler condition for complex Finsler metrics. The Euler-Lagrange equation for a geodesic is

$$\frac{d}{dt} G_{\overline{\beta}}(\dot{x},x) = G_{x\overline{\beta}}(\dot{x},x) . \tag{15}$$

If a holomorphic curve, i.e., the image of disk, is to be geodesic in each radial direction at (ξ,x), we must have both

$$\dot{\xi}^\alpha = -\Gamma^\alpha_\gamma \xi^\gamma \tag{16}$$

and

$$[G_{\alpha\varepsilon}\Gamma^\alpha_\gamma - G_{\varepsilon x^\gamma} + G_{\gamma x^\varepsilon}]\xi^\gamma = 0 . \tag{K_2}$$

This condition assumes the metric C^2 and semi-regular, i.e., $G_{\alpha\overline{\beta}}$ positive definite. If the metric is also of class C^3, the condition K_2 may be rewritten as

$$G_\lambda[\Gamma^\lambda_{\varepsilon,\gamma} - \Gamma^\lambda_{\gamma,\varepsilon}]\xi^\gamma = 0 . \tag{K_3}$$

In semi-normal coordinates for G of class C^3 and semi-regular we have

$$G_{\overline{\beta}\varepsilon x^\gamma}\xi^\varepsilon\xi^\gamma = G_{\overline{\beta}x^\gamma}\xi^\gamma = 0 . \tag{17}$$

If the metric also satisfies K_3 we also have

$$0 = G_{\varepsilon x^\gamma}\,\xi^\gamma \tag{18}$$

and

$$0 = G_{x^\gamma}\ .$$

There is also a form of the Kahler condition for metrics which are only C^1. It is the condition that for each (ξ,x) there are "normal" coordinates so that

$$\frac{d}{dt}\,[G_{\overline{\beta}}(e^{i\theta}\xi, x+te^{i\theta}) = G_{x\overline{\beta}}(e^{i\theta}\xi, x) \tag{K_1}$$

at $t = 0$.

This condition is satisfied if there is a holomorphic map φ of a small disk into M such that $\varphi(0) = x$, $\dot\varphi(0) = \xi$ and $\varphi(e^{i\theta}t)$ geodesic at x for each θ. An example of a Finsler metric of class C' satisfying this condition is the Teichmuller metric.

THEOREM. Let F be a complex Finsler metric of class C' which satisfies K_1 at (ξ,x). Let φ be a map of the disk into M with $\varphi(0) = x$, $\dot\varphi(0) = \xi$. Then the metric F restricted to the normal coordinate line through (ξ,x) is a supporting metric for $F\circ\varphi$. Thus the holomorphic sectional curvature of F at (ξ,x) is the Gauss curvature of F restricted to the normal coordinate line through (ξ,x).

DEPARTMENT OF MATHEMATICS
STANFORD UNIVERSITY
STANFORD, CA 94305

Contemporary Mathematics
Volume 49, 1986

APPLICATIONS OF HARMONIC MAPS TO KÄHLER GEOMETRY

J. H. Sampson[1]

1. INTRODUCTION

Harmonic maps have found many applications to Riemann surfaces because of the intimate connexion between analytic functions and harmonic functions. In higher dimensions a close connexion still persists between harmonic objects and function theory on Kähler manifolds, but it is much more difficult to exploit. However, a certain rigidity which often accompanies the Kähler structure tends to narrow the gap between harmonic and analytic (cf. §2 below).

As an example, it was shown in [E-S, §2] that holomorphic maps of Kähler manifolds are harmonic and that an immersion of riemannian manifolds is minimal if and only if it is harmonic, whence there follows at once that *a complex submanifold of a Kähler manifold is a minimal variety*.

There are many results concerning the existence of minimal immersions of closed Riemann surfaces in various riemannian manifolds (cf. [Bom], [Cal], [Oss]. Here we are going to show that there are severe obstructions to minimal immersions of closed Kähler manifolds of complex dimension > 1 in non-complex spaces of non-positive curvature. The method is a real version of our earlier proof of a theorem of Siu, discussed in §§2, 3. We show in §4 that a harmonic map $f : M \to Y$ of a compact Kähler manifold into a locally symmetric space having no compact local factors gives rise to an abelian Lie algebra of matrices. Such algebras seem to be rather difficult to deal with in general. We illustrate the application when Y is a discrete quotient of $SO_o(p,q)/SO_p \times SO_q$ for $p = 1, 2$. We find that the rank of f is at most 2 when $p = 1$ and is $\leq 2q$ for $p = 2$. The simplest result, treated in [Sa 3] by a different method, is that *a complex Kähler manifold of dimension > 1 cannot be minimally immersed in a space of constant negative curvature*.

1980 Mathematics Subject Classification. 32C10, 53A10, 53C35, 58E20

[1]This work was supported in part by the Consiglio Nazionale delle Ricerche, Comitato per le Scienze Matematiche.

0271-4132/86 $1.00 + $.25 per page

We first comment on this assertion. A Kähler manifold carries with it (1) a complex structure and (2) a riemannian structure. It is readily seen that (1) and (2), but not (2) alone, in turn determine the Kähler structure. Hence the theorem states that a compact riemannian manifold which underlies a Kähler structure of $\dim_{\mathbb{C}} > 1$ cannot be minimally immersed in a space of constant negative curvature. Being minimally immersed or not refers only to the riemannian structure.

It is perhaps germain to recall here a theorem of Bochner [Bo]: A Kähler manifold of dimension n greater than one cannot have constant curvature $\neq 0$. Bochner's proof depends upon a formal computation. However we can use the following geometric argument: Let Ψ_P denote the local holonomy group at a point P of the space. If the space has constant curvature $\neq 0$, then $\Psi_P = SO_{2n}$, whereas the holonomy group of a Kähler n-manifold is a subgroup of SU_n (cf. [K-N, Ch. X]). Of course SU_n is a proper subgroup of SO_{2n} for $n > 1$. The same argument applies to other G-structures as well.

Our results described above can be considered as a kind of generalization of Bochner's, although his is local, whereas ours are global.

2. MAPS OF KÄHLER MANIFOLDS; SIU'S RIGIDITY THEOREM

Here we shall set up the relevant formalism for Kähler manifolds, which we shall then employ to give our very short proof of a fundamental result of Siu. Our discussion in §3 will follow along very similar lines, up to a point. Siu shows that, under suitable hypotheses on curvature and non-degeneracy, a harmonic map of compact Kähler manifolds of the same dimension is necessarily holomorphic (or anti-holomorphic). This depends upon showing that a certain curvature expression vanishes. Cf. [Siu 1, 2].

Let M and M' be Kähler manifolds with metrics denoted by $ds^2 = g_{\alpha\bar\beta} dz^\alpha \, d\bar z^\beta$ resp. $ds'^2 = g'_{a\bar b} dw^a \, d\bar w^b$, referred to local complex coordinate systems. We write $\Gamma^\alpha_{\beta\gamma}$ and Γ'^a_{bc} for the corresponding complex Christoffel symbols; and we recall that in Kähler metric mixed symbols such as $\Gamma^\alpha_{\beta\bar\gamma}$, $\Gamma^{\bar\alpha}_{\beta\gamma}$, etc. all vanish (cf. [Bo], [K-N, Ch.IX]). We write also ∂_α for $\partial/\partial z^\alpha$ and $\partial_{\bar\alpha}$ for $\partial/\partial \bar z^\alpha$.

Let $f : M \to M'$ be a differentiable mapping. Then from the complexifications of the tensor bundles $T^{0,p}(M)$ and $T(M')$ we obtain the vector bundle

$$E_p = T^{0,p}(M)^{\mathbb{C}} \otimes f^{-1} T(M')^{\mathbb{C}} \tag{1}$$

over M. In a standard way, the Kähler connexions on M and M' define a connexion D on E_p, and D splits into $D' + D''$, where D' is of type $(1,0)$ and D'' is of type $(0,1)$. Now the $(1,0)$-part $d'f$ of the differential df is a section of E_1. Its components relative to suitably situated coordinate charts are the derivatives $w^a_\alpha = \partial_\alpha w^a$. The covariant derivatives $D'd'f$ and $D''d'f$ have components

$$(2)\qquad \begin{cases} w^a_{\alpha|\beta} = w^a_{\alpha,\beta} + \Gamma'^a_{bc}\, w^b_\alpha\, w^c_\beta \\[2ex] w^a_{\alpha|\bar\beta} = w^a_{\alpha,\bar\beta} + \Gamma'^a_{bc}\, w^b_\alpha\, w^c_{\bar\beta} \end{cases},$$

where $w^a_{\alpha,\beta}$ and $w^a_{\alpha,\bar\beta}$ denote covariant derivatives with respect to the connexion on M. Thus $w^a_{\alpha,\beta} = \partial_\beta w^a_\alpha - \Gamma^\gamma_{\alpha\beta}\, w^a_\gamma$ and $w^a_{\alpha,\bar\beta} = \partial_{\bar\beta}\, w^a_\alpha$. The derivatives $D'D'd'f$, $D''D'd'f$, etc. are formed following the pattern of (2). For example, $D''D''d'f$ has components

$$w^a_{\alpha|\bar\beta|\bar\gamma} = \partial_{\bar\gamma} w^a_{\alpha|\bar\beta} - \Gamma^{\bar\mu}_{\bar\beta\bar\gamma}\, w^a_{\alpha|\bar\mu} + \Gamma'^a_{bc}\, w^b_{\alpha|\bar\beta}\, w^c_{\bar\gamma}\ .$$

From this we easily obtain the following relation:

$$(3)\qquad w^a_{\alpha|\bar\beta|\bar\gamma} - w^a_{\alpha|\bar\gamma|\bar\beta} = R'^a_{b\bar c d}\, w^b_\alpha (\bar w^c_{\bar\gamma} w^d_{\bar\beta} - \bar w^c_{\bar\beta} w^d_{\bar\gamma})\ ,$$

the $R'^a_{b\bar c d}$ being the components of the curvature tensor of M'. We observe that (3) does not contain the curvature tensor of M, a fundamental point in what follows.

The harmonic equation for f is simply

$$(4)\qquad g^{\alpha\bar\beta}\, w^a_{\alpha|\bar\beta} = 0,$$

from which it is clear that (4) vanishes if f is holomorphic (the proof in [E-S] is quite different).

Let us now introduce the covariant tensor ϕ of type $(2,0)$ with components

$$(5)\qquad \phi_{\alpha\beta} = g'_{a\bar b}\, w^a_\alpha\, \bar w^b_\beta\ ,$$

which is the $(2,0)$-part of the pull-back $f^*(ds'^2)$. As in [Sa 2] we pass to the $(1,0)$ form ξ with components

$$(6)\qquad \xi_\alpha = g^{\beta\bar\gamma}\, \phi_{\alpha\beta,\bar\gamma} \qquad \text{(covariant derivative)}.$$

We now assume that the mapping f is harmonic. From (4) we obtain

$$\xi_\alpha = g'_{a\bar b}\, w^a_{\alpha|\bar\gamma}\, \bar w^b_\beta\, g^{\beta\bar\gamma}\ .$$

The divergence of ξ is the scalar $\delta\xi = g^{\alpha\bar{\mu}}\xi_{\alpha,\bar{\mu}}$. Using (3) we get

$$\delta\xi = g'_{a\bar{b}} w^a_{\alpha|\bar{\gamma}} \overline{w}^b_{\beta|\bar{\mu}} g^{\alpha\bar{\mu}} g^{\beta\bar{\gamma}} \tag{7}$$
$$+ R'_{a\bar{b}c\bar{d}} \overline{w}^a_{\beta} w^b_{\alpha} (\overline{w}^c_{\bar{\mu}} w^d_{\bar{\gamma}} - \overline{w}^c_{\bar{\gamma}} w^d_{\bar{\mu}}) g^{\alpha\bar{\mu}} g^{\beta\bar{\gamma}} .$$

The first term on the right is easily seen to be ≥ 0. By a simple transformation the curvature term can be written

$$\frac{1}{2} R'_{a\bar{b}c\bar{d}} (\overline{w}^a_{\beta} w^b_{\alpha} - \overline{w}^a_{\alpha} w^b_{\beta}) \overline{(\overline{w}^d_{\gamma} w^c_{\mu} - \overline{w}^d_{\mu} w^c_{\gamma})} g^{\alpha\bar{\mu}} g^{\beta\bar{\gamma}} . \tag{8}$$

Under the hypothesis that the curvature of M' is <u>strongly semi-negative</u>, as defined in [Siu 1], we have $\delta\xi \geq 0$. If M is compact, then from the divergence theorem there follows $\delta\xi = 0$. Consequently both terms on the right of (7) must vanish. In particular, all $w^a_{\alpha|\bar{\beta}}$ are zero, and so $\phi_{\alpha\beta,\bar{\gamma}} = 0$. I.e., <u>the 2-tensor ϕ is holomorphic</u>. It is the vanishing of (8) that allows Siu, under appropriate conditions, to conclude that all w^a_{α} or else all $w^a_{\bar{\alpha}}$ must be zero.

3. MAPS FROM KÄHLER TO RIEMANNIAN

We now embark upon a similar analysis for a mapping $f : M \to Y$ of a Kähler manifold M into a riemannian manifold Y. For M we retain the foregoing notation. The metric tensor of Y will be written

$$ds'^2 = g_{jk} dy^j dy^k ,$$

with corresponding Christoffel symbols Γ^i_{jk} and Riemann tensor R^i_{jkl}.

The pull-back $f^*(ds'^2)$ decomposes into three terms, $\phi + \psi + \bar{\phi}$, of types (2,0), (1,1), (0,2) respectively. In particular ϕ has components

$$\phi_{\alpha\beta} = g_{jk} y^j_{\alpha} y^k_{\beta} , \tag{9}$$

where we write in general $u_\alpha = \partial_\alpha u$ and $u_{\bar{\alpha}} = \partial_{\bar{\alpha}} u$. Observe that, for real u, we have $u_{\bar{\alpha}} = \overline{u_{\alpha}}$. We note that ϕ is symmetric, $\phi_{\alpha\beta} = \phi_{\beta\alpha}$, which was not necessarily the case with the tensor ϕ of §2. With ϕ we associate the 1-form ξ whose components are

$$\xi_{\alpha} = g^{\beta\bar{\gamma}} \phi_{\alpha\beta,\bar{\gamma}} . \tag{10}$$

We shall again compute the divergence $\delta\xi = g^{\alpha\bar{\mu}}\xi_{\alpha,\bar{\mu}}$. In the present situation the bundles E_p of (1) are replaced by the vector bundles

(11) $$F_p = T^{0,p}(M)^{\mathbb{C}} \otimes T(Y)^{\mathbb{C}}.$$

Covariant derivatives D are introduced by $\mathbb{C}$-linear extension. The harmonic mapping equation then takes the form

(12) $$g^{\alpha\bar{\beta}} y^j_{\alpha|\bar{\beta}} = 0,$$

where

$$y^j_{\alpha|\bar{\beta}} = \partial_{\bar{\beta}} y^j_\alpha + \Gamma^j_{kl}\, y^k_\alpha y^l_{\bar{\beta}} .$$

If f is harmonic there follows

$$\xi_\alpha = g_{jk}\, y^j_{\alpha|\bar{\gamma}}\, y^k_\beta\, g^{\beta\bar{\gamma}} .$$

Then

$$\xi_{\alpha,\bar{\mu}} = g_{jk}(y^j_{\alpha|\bar{\gamma}|\bar{\mu}}\, y^k_\beta + y^j_{\alpha|\bar{\gamma}}\, y^k_{\beta|\bar{\mu}}) g^{\beta\bar{\gamma}} .$$

The Ricci identity (3) is replaced by the analogous relation

$$y^j_{\alpha|\bar{\gamma}|\bar{\mu}} - y^j_{\alpha|\bar{\mu}|\bar{\gamma}} = - R^j_{klm}\, y^k_\alpha y^l_{\bar{\gamma}} y^m_{\bar{\mu}} .$$

There follows easily

(13) $$\begin{aligned} \delta\xi &= g_{jk} y^j_{\bar{\alpha}|\gamma} y^k_{\beta|\bar{\mu}}\, g^{\beta\bar{\gamma}} g^{\alpha\bar{\mu}} \\ &\quad - R_{jklm} y^j_\beta y^k_\alpha y^l_{\bar{\gamma}} y^m_{\bar{\mu}} g^{\beta\bar{\gamma}} g^{\alpha\bar{\mu}} . \end{aligned}$$

It is readily seen that the curvature term is real. The other term on the right of (13) is ≥ 0.

<u>DEFINITION</u>. The curvature of Y will be called <u>hermitian negative</u> if

$$R_{jklm} u^j v^k \bar{u}^l \bar{v}^m \leq 0$$

for arbitrary complex vectors u and v.

If that is the case, then we have, as in [Sa 3]:

<u>THEOREM</u> 1. If f is a harmonic map of a compact Kähler manifold into a riemannian manifold of hermitian negative curvature, then ϕ is holomorphic, and we have

(14) $$y^j_{\alpha|\bar{\beta}} = 0;\ R_{jklm} y^j_\beta y^k_\alpha y^l_{\bar{\gamma}} y^m_{\bar{\mu}} g^{\beta\bar{\gamma}} g^{\alpha\bar{\mu}} = 0.$$

Indeed, our assumptions imply that $\delta\xi \geq 0$. The rest follows at once from the divergence theorem and (13). We observe that if $\dim_{\mathbb{C}} M = 1$, then the harmonic equation reduces to $y^j_{1|\bar{1}} = 0$, whence $\phi_{11,\bar{1}} = 0$. Thus ϕ is holomorphic without curvature or compactness assumptions, as is well known.

4. CURVATURE IN LOCALLY SYMMETRIC SPACES

We now turn to some examples of spaces to which Theorem 1 can be applied. A riemannian locally symmetric space has the form $Y = \Gamma\backslash G/K$, where K is a connected and closed subgroup of the connected Lie group G and where Γ is a discrete subgroup of G. Curvature computations which we shall require are purely local and G-invariant, and therefore it suffices to make them at the origin o of the homogeneous space G/K. Let the corresponding Cartan decomposition of the Lie algebra of G be

$$\underline{g} = \underline{k} + \underline{p} .$$

Then the subspace $\underline{p}$ can be identified with the tangent space of G/K at o. The curvature tensor of G/K at o corresponding to any G-invariant metric is given by Cartan's formula

$$R(X,Y)Z = -[[X,Y],Z] \quad \text{for} \quad X, Y, Z \text{ in } \underline{p} .$$

(Cf. [Hel, Ch. IV] and [K-N, Ch. XI]. Therefore, if X_i denotes an orthonormal basis for the given riemannian structure, then we have

$$R_{jklm} = ([[X_j,X_k],X_l], X_m). \tag{15}$$

In all that follows we shall be concerned solely with linear algebras $\underline{g}$, and therefore we take the elements of $\underline{g}$ to be square matrices. Since we must deal with the complexification of $\underline{g}$, or at least of $\underline{p}$, it will be prudent to assume that the elements of $\underline{g}$ are real. That is no loss; for if $\underline{g}$ contains complex matrices $X = X' + iX''$ (X' and X'' real), then the operation $X \longrightarrow \begin{pmatrix} X' & X'' \\ -X'' & X' \end{pmatrix}$ maps $\underline{g}$ isomorphically upon an algebra of real matrices.

Suppose now that $\underline{g}$ is a simple algebra of non-compact type, and consider the trace form $(X,Y) = \mathrm{tr}({}^tXY)$ on $\underline{g}$. It defines a riemannian metric on G/K, being positive definite on $\underline{p}$ and negative definite on $\underline{k}$. Because of its invariance under the adjoint action, (15) can now be written

$$R_{jklm} = ([X_j,X_k],[X_l,X_m]) \tag{16}$$

(cf. [Hel, Ch. V, §3] and [Wo, Ch. 8, §4]). If r is the dimension of $\underline{p}$, let $u\ (u^1,\ldots,u^r)$ and $v = (v^1,\ldots,v^r)$ be complex vectors, and set

$X = u^j X_j$, $Y = v^j X_j$. Then we have at once from (16), recalling that $[\underline{p},\underline{p}] \subset \underline{k}$,

$$R_{jklm} u^j v^k \bar{u}^l \bar{v}^m = ([X,Y],[\bar{X},\bar{Y}]), \tag{17}$$

which is clearly ≤ 0. This also holds trivially for a symmetric space of euclidean type, since there the curvature tensor is zero. Hence, referring to the Definition of §2, we have

THEOREM 2. Let Y be a riemannian locally symmetric space whose irreducible local factors are all of the non-compact or euclidean type. Then Y has hermitian negative curvature.

Let us now suppose that the second equation of (14) holds. Choosing coordinates z^α at a point $P \in M$ such that $g_{\alpha\bar{\beta}} = \delta_{\alpha\beta}$ there, we have, for each pair α, β,

$$R_{jklm} y^j_\beta y^k_\alpha \bar{y}^l_\beta \bar{y}^m_\alpha = 0,$$

and therefore from (17) there follows

$$[Z_\alpha, Z_\beta] = 0 \quad \text{for all} \quad Z_\alpha = y^j_\alpha X_j . \tag{18}$$

The complex tangent space $T_P(M)^C$ splits into $T'_P(M) + T''_P(M)$, of types (1,0) and (0,1) respectively. The former is spanned by the operators $\partial_1, \ldots, \partial_m$ ($m = \dim_{\mathbb{C}} M$). It is mapped by the $\mathbb{C}$-linear part $d'f$ of df onto the space $\underline{a} \subset \underline{p}^{\mathbb{C}}$ spanned by the elements (18). Here of course we have identified $T_{f(P)}(Y)$ with $\underline{p}$, as above. Applying this to the local irreducible factors of Y, we obtain

THEOREM 3. Under the conditions of Theorem 2, if the second equation of (14) holds, then $d'f$ maps $T'_P(M)$ onto an abelian algebra $\underline{a} \subset \underline{p}^{\mathbb{C}}$, where $\underline{p}$ is identified with $T_{f(P)}(Y)$. The hypothesis concerning eq. (14) is satisfied if f is harmonic and M is compact.

4. ABELIAN ALGEBRAS

Theorem 3 implies some rather severe restriction upon harmonic maps of compact Kähler manifolds into locally symmetric spaces. However we cannot expect simple and sweeping statements. We first discuss some examples.

EXAMPLE. Let $\underline{g} = \underline{sp}_n \mathbb{R}$, consisting of all matrices

$$\begin{pmatrix} X_1 & X_2 \\ X_3 & -{}^tX_1 \end{pmatrix},$$

where X_1, X_2, X_3 are real $n \times n$ matrices, with X_2 and X_3 symmetric.

We have the Cartan decomposition $\underline{g} = \underline{k} + \underline{p}$, where $\underline{k}$ consists of all $\begin{pmatrix} A & B \\ -B & -A \end{pmatrix}$, A skew-symmetric and B symmetric. By mapping the matrix to $A + iB$, we can identify $\underline{k}$ with $\underline{u}_n$. The subspace $\underline{p}$ consists of all $\begin{pmatrix} U & V \\ V & -U \end{pmatrix}$, with symmetric U and V. The element $J_o = \begin{pmatrix} 0 & I \\ -I & 0 \end{pmatrix}$ is in the center of $\underline{k}$, and conjugation by $e^{\pi J_o/4}$ defines a complex structure J on $\underline{p}$, with $J : \begin{pmatrix} U & V \\ V & -U \end{pmatrix} \to \begin{pmatrix} V & -U \\ -U & -V \end{pmatrix}$. Thus $\underline{k} + \underline{p}$ is the underlying orthogonal Lie algebra of a hermitian symmetric space. Matrices of the form

$\begin{pmatrix} U & iU \\ V & -U \end{pmatrix}$ with symmetric U have all products $= 0$ (hence are nilpotent). They constitute an abelian Lie algebra $\underline{a} \subset \underline{p}^{\mathbb{C}}$ of real dimension $n^2 + n$, which is the same as the dimension of $\underline{p}$. The rank in this case is just n. Hence we see that there is no close connexion between $\dim \underline{a}$ and the rank.

Now $\underline{sp}_n \mathbb{R}$ is naturally embedded in $\tilde{\underline{g}} = \tilde{\underline{k}} + \tilde{\underline{p}}$, where $\tilde{\underline{g}} = \underline{sl}_{2n} \mathbb{R}$, $\tilde{\underline{k}} = \underline{so}_{2n}$. In this case $\tilde{\underline{p}}$ consists of all real $2n \times 2n$ matrices of trace zero. It contains the preceding $\underline{p}$, and $\tilde{\underline{k}}$ contains the preceding $\underline{k}$. Thus $\tilde{\underline{p}}^{\mathbb{C}}$ contains abelian Lie algebras of real dimension $n^2 + n$. We do not know whether or not there are bigger ones. As for the symmetric spaces, we have

<u>THEOREM</u> 4. The natural embedding of $Sp_n \mathbb{R}/U_n$ is $SL_{2n} \mathbb{R}/SO_{2n}$ is totally geodesic.

We recall (cf. [E-S, §2] and [Sa 1]) that if we have maps of riemannian manifolds $X \xrightarrow{f} Y \xrightarrow{g} Z$ with f harmonic and g totally geodesic, then $g \circ f$ is harmonic. Hence a harmonic map into $Sp_n \mathbb{R}/U_n$ (or a discrete quotient thereof) is also harmonic as viewed in $SL_{2n} \mathbb{R}/SO_{2n}$.

The problem of determining maximum dimensions of abelian algebras $\underline{a} \subset \underline{p}^{\mathbb{C}}$ seems to be rather difficult in general. Here we shall limit our considerations to special case of symmetric spaces of type BD I: $SO_o(p,q)/SO_p \times SO_q$, $p \leq q$. Here $\underline{g} = \underline{k} + \underline{p}$, where $\underline{k}$ consists of all matrices $\begin{pmatrix} A & 0 \\ 0 & D \end{pmatrix}$, $A \in \underline{so}_p$ and $D \in \underline{so}_q$; $\underline{p}$ consists of all $\begin{pmatrix} 0 & B \\ {}^tB & 0 \end{pmatrix}$, B of size $p \times q$. The dimension is pq, and the rank is p. Let $\underline{a} \subset \underline{p}^{\mathbb{C}}$ be an abelian Lie algebra, and take elements $\alpha = \begin{pmatrix} 0 & A \\ {}^tA & 0 \end{pmatrix}$, $\beta = \begin{pmatrix} 0 & B \\ {}^tB & 0 \end{pmatrix}$ in $\underline{a}$. The condition $[\alpha,\beta] = 0$ becomes $A\,{}^tB = B\,{}^tA$ and ${}^tAB = {}^tBA$. If $p = 1$

we have $A = (a_1,\ldots,a_q)$ and $B = (b_1,\ldots,b_q)$, and $a_i b_j = a_j b_i$. We have therefore

PROPOSITION 1. If $p = 1$, a maximal abelian algebra in $\underline{p}^{\mathbb{C}}$ has $\text{dimension}_{\mathbb{C}} = 1$.

We next consider $p = 2$. In this case we have $a_{1i}b_{1j} + a_{2i}b_{2j} = b_{1i}a_{1j} + b_{2i}a_{2j}$. Suppose that some matrix A has rank 2. We may assume for example that the first two columns of A are linearly independent. We see at once that all the elements b_{ij} of B are determined by b_{11}, b_{12}, b_{21}, b_{22}. It is not difficult to show from this that $\dim_{\mathbb{C}} \underline{a} \leq 2$. If no matrix A in our collection has rank = 2, then the two rows of any A are proportional. In this situation the maximum dimension of $\underline{a}$ is q, with $\underline{a}$ comprised of matrices $\begin{pmatrix} a_1 \ldots a_q \\ ia_1 \ldots ia_q \end{pmatrix}$ or $\begin{pmatrix} a_1 \ldots a_q \\ -ia_1 \ldots -ia_q \end{pmatrix}$ (but not both).

PROPOSITION 2. For $p = 2$ an abelian Lie algebra in $\underline{p}^{\mathbb{C}}$ has dimension $\leq q$.

We shall not carry this line of investigation further here. From the foregoing and Theorem 3 we can now state

THEOREM 5. A harmonic map of a compact Kähler manifold into a locally symmetric space of type BD I (p,q) has rank ≤ 2 if $p = 1$, and $\leq 2q$ if $p = 2$, $q \geq 2$.

The case $p = 1$ is of course that of constant negative curvature, treated by a quite different method in [Sa 3].

REFERENCES

[Bo] S. Bochner, Curvature in hermitian metric, Bull. Amer. Math. Soc. 53 (1947), 179-195.

[Bom] E. Bombieri, Seminar on Minimal Submanifolds, Annals of Math. Studies 103, Princeton (1983).

[Cal] E. Calabi, Minimal immersions of surfaces in euclidean spheres, J. Diff. Geom. 1 (1967), 111-125.

[E-S] J. Eells and J. H. Sampson, Harmonic mappings of riemannian manifolds, Amer. Jour. Math. 86 (1964), 109-160.

[Hel] S. Helgason, Differential Geoemtry and Symmetric Spaces, Academic Press, New York (1962).

[K-N] S. Kobayashi and K. Nomizu, Foundations of Differential Geometry, Interscience, New York (1969).

[Oss] A Survey of Minimal Surfaces, Van Nostrand Reinhold, New York 1969.

[Sa 1] On a class of harmonic maps, Harmonic Map Proceedings, Springer Lecture Notes 949 (1982), 138-139.

[Sa 2] On harmonic mappings, Symp. Math. XXVI (1982), 197-210.

[Sa 3] Harmonic mappings and minimal immersions, C.I.M.E. Corso Estivo on Harmonic Mappings and Minimal Immersions, Montecatini T. (1984).

[Siu 1] Complex analyticity of harmonic maps and strong rigidity of complex Kähler manifolds, Ann. Math. 112 (1980), 73-111.

[Siu 2] Complex analyticity of harmonic maps, vanishing and Lefschetz theorems, J. Diff. Geom. 17 (1982), 55-138.

[Wo] J. Wolf, Spaces of Constant Curvature, Publish or Perish, Boston (1974).

DEPARTMENT OF MATHEMATICS
THE JOHNS HOPKINS UNIVERSITY
BALTIMORE, MARYLAND 21218

Contemporary Mathematics
Volume **49**, 1986

ON THE RELATION BETWEEN CHERN AND PONTRJAGIN NUMBERS

S.M. Webster*

ABSTRACT. We consider a real n-plane subbundle V of a complex n-plane bundle $\tilde{V}$ over an oriented real n-manifold. For $n=4$, we derive a formula for the Pontrjagin number of V which is useful in studying hulls of holomorphy. For $n=2$, we derive a formula for the first Chern number of $\tilde{V}$ and apply it to the study of minimal surfaces in $\mathbb{P}_2\mathbb{C}$.

If V is a real vector bundle over a manifold, then its k-th Pontrjagin class is related to the 2k-th Chern class of its complexification V^c by $p_k(V) = (-1)^k c_{2k}(V^c)$. This formula is in fact often taken as the definition of $p_k(V)$, since the Chern classes are simpler invariants. V is a totally real subbundle of V^c of half its real rank. In this paper we consider a complex vector bundle $\tilde{V}$ of rank n and a real subbundle $V \subset \tilde{V}$ of real rank n over a compact, oriented (real) n-manifold M. We denote by $J, J^2 = -I$, the linear operator on the underlying real 2n-plane bundle of $\tilde{V}$ given by multiplication by $i = \sqrt{-1}$. Let

$$H_p = V_p \cap JV_p \ , \quad N = \{p \in M \colon \dim V_p > 0\} \ .$$

N is the set of points over which the fiber of V contains a non-trivial complex subspace. If $n=4$ and V is a generic real subbundle of $\tilde{V}$, then it can be shown [4] that $\dim_{\mathbb{C}} H_p \leq 1$, and N is a disjoint union of compact, connected, naturally oriented surfaces. In this case the spaces H_p form a complex line bundle $H \to N$. Our main result is the following.

1980 Mathematics Subject Classification. 53C99
*Partially supported by NSF Grant No. MCS-8300245.

Theorem 1) Let V be a generic real 4-plane subbundle of the complex 4-plane bundle $\tilde{V}$ over a compact, oriented 4-manifold M. Then the Chern and Pontrjagin numbers satisfy

$$(1) \qquad c_2(\tilde{V})[M] = -p_1(V)[M] + c_1(H)[N] .$$

In the special case where M is generically immerse in $\mathbb{C}^4$, $V = TM$, and $\tilde{V} = M \times \mathbb{C}^4$, the relation (1) was derived in [4]. We also consider the simpler case of 2-plane bundles over a compact, oriented surface. In case N is a discrete set, we have assigned in [5] an index to each point of N. Here we give a formula relating the sum of these indices to the first Chern number of $\tilde{V}$. This has applications to minimal surfaces embedded in the complex projective plane.

It is most convenient to study $V \subset \tilde{V}$ via their complexifications $V^c \subset \tilde{V}^c$;

$$\tilde{V}^c = V' \oplus V'' , \quad V'' = \overline{V'} , \quad J = iI \text{ on } V' ,$$

$$H^c = H' \oplus H'' , \quad H'' = \overline{H'} , \quad H' = H^c \cap V' .$$

Let $\pi' : \tilde{V}^c \to V'$ be the complex-linear projection along V''. π'_p induces a linear isomorphism $V^c_p \to V'_p$, if $p \notin N$. If $p \in N$, π'_p has null space H''_p and is the identity on H'_p. On the n-th exterior powers π' induces a mapping of complex line bundles

$$\Lambda^n \pi' : \Lambda^n(V^c) \to \Lambda^n(V') .$$

which we may regard as a smooth section of the complex line bundle

$$S \equiv \Lambda^n(V^c)^* \otimes \Lambda^n(V') .$$

It is clear that N is precisely the zero set of this section. Also, the (n-2)-cycle determined by N is dual to $c_1(S)$, which equals $c_1(\tilde{V})$ modulo torsion, since $\pi' : \tilde{V} \cong V'$, and $2c_1(V^c) = 0$ (see [2]).

Locally, we choose frame fields $(e_1, \ldots, e_n)$ in V and $(v_1, \ldots, v_n)$ in V'. Then

$$\pi'(e_j) = \frac{1}{2}(e_j - iJe_j) ,$$

(2) $$\pi'(e_1 \wedge \ldots \wedge e_n) = 2^{-n}(e_1 - iJe_1) \wedge \ldots \wedge (e_n - iJe_n)$$
$$= Bv_1 \wedge \ldots \wedge v_n ,$$

for a smooth complex valued function B. In case the non-degeneracy condition

(3) $$dB \wedge d\bar{B} \neq 0 \text{ on } N$$

holds, as it does generically for $n < 8$ [4], N is a smooth submanifold of codimension 2 in M. Its normal bundle is the restriction of the complex line bundle S to N. It follows that N has a naturally induced orientation from that of M.

We recall the definition of Chern classes of a complex n-plane bundle $F \to M^n$ via obstruction theory [1], [3]. Let $St(r,n)$ denote the Stiefel manifold of unitary r-frames $f = (f_1,\ldots,f_r)$ in $\mathbb{C}^n$. If we fix $f_1,\ldots,f_{r-1}$ and vary f_r, we get a (2n-2r+1)-sphere embedded in $St(r,n)$. It is the canonical generator of the first non-trivial homotopy group of $St(r,n)$. Let $M(k)$ denote the k-skeleton of a smooth triangulation of M,

$$M(0) \subset \ldots \subset M(k) \subset \ldots \subset M(n) = M ,$$

which is so fine that F is trivial over each simplex. An r-frame f in F may be chosen over each vertex of $M(0)$ and extended to $M(j)$ for $j \leq 2n - 2r + 1$. For each simplex Δ_{2k} of dimension $2k = 2n - 2r + 2$, f gives a map from $\partial\Delta_{2k}$ into $St(r,n)$, i.e. an element $f(\Delta)$ of $\pi_{2k-1}(St(r,n)) \cong \mathbb{Z}$. The assignment $\Delta \to f(\Delta)$ is a 2k-cocycle on M with integral coefficients representing $c_k(F)$.

Given the smooth submanifold $N^{n-2} \subset M^n$, we may find a triangulation which is transverse to N. In fact we may choose a diffeomorphism φ of M, arbitrarily close to the identity in some C^k sense, so that $\varphi(N)$ is transverse to every open simplex of some given smooth triangulation $M_0(k)$ of M.

We then replace $M_0(k)$ by $\varphi^{-1}M_0(k)$. This induces a triangulation $N(k-2) = N \cap M(k)$, $2 \leq k \leq n$, of N . We may assume that $N(n-3)$ is disjoint from some given finite subset of N , and that each simplex of the triangulation has diameter less than $\epsilon_0 > 0$ to be prescribed below.

We return to the situation of Theorem 1). The idea of its proof is to relate $c_2(V^c)$ and $c_2(V')$ via the map π' . We choose a smooth hermitian metric g in the fibers of $\tilde{V}$ and utilize the real metric $\mathcal{R}eg$. Let $W \to N$ be the totally real 2-plane bundle which is the orthogonal complement of H in V . Modifying g if necessary, we may assume that W and JW are orthogonal along N . The complexification $W^c \to N$ is a trivial complex 2-plane bundle, as may easily be seen as follows. Since $2c_1(W^c) = 0$ in the torsion free group $H^2(N,\mathbb{Z})$, we have $c_1(W^c) = 0$. Hence, $\Lambda^2 W^c$ is trivial, meaning that W^c admits a global non-degenerate 2-form φ . W^c also admits a never-zero section $f_1 = f_1' + if_1''$, where f_1', f_1'' are sections of W with disjoint zero sets. A second independent section f_2 of W^c is determined by requiring $\varphi(f_1, f_2) = +1$, and $f_2 \perp f_1$ relative to some hermitian metric on W^c . Locally we may think of the frame (f_1, f_2) as a map from N to $G\ell(2,\mathbb{C})$, which is connected. After a smooth deformation we may assume that (f_1, f_2) take prescribed values in a suitable small neighborhood of a given point of N . Next we choose sections w' of H' and w'' of H'' over N , which have disjoint, discrete zero sets. If we set $f_3 = w' + w''$, then $f = (f_1, f_2, f_3)$ is a 3-frame in V^c along N . We extend f smoothly to a 3-frame in V^c over a neighborhood $\mathcal{U}$ of N in M . $\tilde{f} = \pi' f$ is a 3-frame in V' over $\mathcal{U}$, except at the finite set of points of N where $\pi' f_3 = w'$ vanishes.

To understand what happens at a point $p_0 \in N$ where $w'(p_0) = 0$, we choose an orthonormal frame e_i , $1 \leq i \leq 8$, in $\tilde{V}$ near p_0 as follows. Along N , e is unitary, $Je_{2\alpha-1} = e_{2\alpha}$, $1 \leq \alpha \leq 4$; e_1, e_2 span H ; e_3, e_5 span W ; e_4, e_6 span JW ; and e_7, e_8 span $(H \oplus W \oplus JW)^\perp$. Such a frame is then extended smoothly to a 4-dimensional neighborhood of p_0 so that

$e_1, e_2, e_3, 3_5$ remain in V. We set

$$E_\alpha = \frac{1}{2}(e_{2\alpha-1} - ie_{2\alpha}) , \quad 1 \leq \alpha \leq 4 .$$

Also, we choose a unitary frame field v_α, $1 \leq \alpha \leq 4$, in V' near p_0, so that $v_\alpha = E_\alpha$ on N. We then have

$$(4) \qquad \begin{aligned} E_\alpha &= a_{\alpha\beta} v_\beta + b_{\alpha\beta} \overline{v}_\beta , \\ a_{\alpha\beta} &= \delta_{\alpha\beta} , \quad b_{\alpha\beta} = 0 \quad \text{on } N . \end{aligned}$$

Following (2) we compute

$$\begin{aligned} \pi'(e_1 \wedge e_2 \wedge e_3 \wedge e_5) &= -2i\pi'\{E_1 \wedge \overline{E}_1 \wedge (E_2 + \overline{E}_2) \wedge (E_3 + \overline{E}_3)\} \\ &= -2i(a_{1\beta} v_\beta) \wedge (\overline{b}_{1\beta} v_\beta) \wedge (a_{2\beta} v_\beta) \wedge (a_{3\beta} v_\beta) + O(\rho^2) \\ &= Bv_1 \wedge v_2 \wedge v_3 \wedge v_4 , \\ B &= -2i(\overline{b}_{14} + O(\rho^2)) , \end{aligned}$$

where ρ is the distance to N. By the non-degeneracy assumption (3), $d\overline{b}_{14} \wedge db_{14} \neq 0$ on N. Thus, properly oriented local coordinates x can be chosen near p_0 by taking $x_3 + ix_4 = \overline{b}_{14}$, and x_1, x_2 local coordinates on N so that $x(p_0) = 0$ and

$$dx_1 \wedge dx_2 \wedge dx_3 \wedge dx_4 = +dV_M .$$

We fix such coordinates and frames for each zero p_0 of w'.

We may assume that w' and w'' are chosen so that near p_0 on N

$$w' = \lambda E_1 , \quad w'' = \overline{E}_1 , \quad \lambda = (x_1 \pm ix_2)^k , \ K > 0 ;$$

the zero of w' at p_0 having index $\pm k$. By our above remarks the frame f may be chosen to satisfy

$$f_1 = e_3 , \quad f_2 = e_5 , \quad f_3 = \lambda E_1 + \overline{E}_1$$

near p_0. By (4) the projected frame $\tilde{f} = \pi' f$ satisfies

$$\tilde{f}_1 = (a_{2\beta} + \overline{b}_{2\beta})v_\beta = v_2 + O(\rho)$$

$$\tilde{f}_2 = (a_{3\beta} + \overline{b}_{3\beta})v_\beta = v_3 + O(\rho)$$

$$\tilde{f}_3 = (\lambda a_{1\beta} + \overline{b}_{1\beta})v_\beta \equiv (\lambda + \overline{b}_{11})v_1 + \overline{b}_{14}v_4 + O(|x|^2) \quad ,$$

$$\mathrm{mod}(v_2,v_3) \quad .$$

Hence, we may find $\epsilon_0 > 0$ such that if $\Delta_4 \subset \{|x| < \epsilon_0\}$, $p_0 \in \mathrm{int}(\Delta_4)$, then as maps from $\partial\Delta_4$ into $St(3,4)$, $\tilde{f}$ and $\hat{f}$ are smoothly homotopic, where

$$\hat{f}_1 = v_2 \ , \quad \hat{f}_2 = v_3 \ , \quad \hat{f}_3 = (\lambda + \overline{b}_{11})v_1 + \overline{b}_{14}v_4 \quad .$$

Since $\overline{b}_{11} = O(\rho)$, $\hat{f}_3$ ahs the coordinate form

$$x \to ((x_1 \pm ix_2)^k + O(|x_3| + |x_4|) \ , \ x_3 + ix_4) \in \mathbb{C}^2 \quad .$$

This maps the circle $x_3 = x_4 = 0$ on $|x| = 1$ around the circle $z_2 = 0$ on $S^3 \subset \mathbb{C}^2$ $\pm k$ times. By the second coordinate the orientations of the normal bundles of these two circles correspond. As the complements of these circles correspond, it follows that $\hat{f}$ determines $\pm k$ times the canonical generator of $\pi_3(St(3,4))$.

Now we determine $\epsilon_0 > 0$ so that the above holds for all zeros of w'. Let $M(k)$ be a triangulation of M transverse to N so that $N \cap M(3)$ is disjoint from the zero set of w' and each simplex has diameter less than ϵ_0. We also assume that $\Delta_4 \subset \mathcal{U}$ if $\Delta_4 \cap N \neq \emptyset$, and that V and $\tilde{V}$ are trivial over each simplex. We extend f, and hence $\tilde{f}$, to $\mathcal{U} \cup M(3)$. By the previous paragraph

$$(5) \qquad \sum_{\Delta_4 \cap N \neq \emptyset} \tilde{f}(\Delta_4) = \sum_{w'(p)=0} \mathrm{ind}_p(w') = c_1(H')[N] \quad .$$

If $\Delta_4 \cap N = \emptyset$, we claim $f(\Delta_4) = \tilde{f}(\Delta_4)$. Over such a Δ_4 π' has the local form $\pi'(p,z) = (p,\varphi(p,z))$ with $\varphi(p,\cdot)$ a linear isomorphism of $\mathbb{C}^4$. We may deform $\varphi(p,\cdot)$ to $\varphi(p_1,\cdot)$, where p_1 is the barycenter of Δ_4, and then deform $\varphi(p_1,\cdot)$ to the identity in $G\ell(4,\mathbb{C})$. This gives a homotopy from f

to $\tilde{f}$ on $\partial\Delta_4$. Since $f(\Delta_4) = 0$ if $\Delta_4 \cap N \neq \emptyset$,

$$(6) \qquad \sum_{\Delta_4 \cap N = \emptyset} \tilde{f}(\Delta_4) = \sum_{\Delta_4} f(\Delta_4) = c_2(V^c)[M] \quad .$$

Combining (5) and (6) we get

$$c_2(V')[M] = c_2(V^c)[M] + c_1(H')[N] \quad .$$

Since $c_2(V') = c_2(\tilde{V})$, $c_1(H') = c_1(H)$, and $c_2(V^c) = -p_1(V)$, Theorem 2) follows.

Next we turn to the case of 2-plane bundles $V \subset \tilde{V}$ over a compact, orientable surface M^2. We assume only that the set N is finite and give M a sufficiently fine triangulation with $N \cap M(1) = \emptyset$. To compute $c_1(V^c)$ we must construct a 2-frame in V^c over $M(1)$ and try to extend it to M. First we choose $f_1(p_0) \in H'_{p_0}$ and $f_2(p_0) \in H''_{p_0}$ and extend $f = (f_1, f_2)$ as a frame in V^c smoothly to the 2-simplex containing p_0, for each p_0 in N. Then we extend f to the 1-skeleton $M(1)$ and define $\tilde{f} = \pi' f$. As before $f(\Delta_2) = 0$ if $N \cap \Delta_2 \neq \emptyset$, and $\tilde{f}(\Delta_2) = f(\Delta_2)$ if $N \cap \Delta_2 = \emptyset$. We choose frames E_α in $\tilde{V}$, v_α in V', $\alpha = 1,2$, near each $p_0 \in N$ as before, with $\mathfrak{Re}\, E_1$, $\mathfrak{Im}\, E_1$ spanning V. As in [5] these frames may be chosen to satisfy

$$E_1 = qv_1 + r\bar{v}_2, \quad q(p_0) = 1, \; r(p_0) = 0,$$

$$E_2 = sv_2 + t\bar{v}_1, \quad s(p_0) = 1, \; t(p_0) = 0 .$$

r vanishes precisely at p_0. Up to a sign the index, $\operatorname{ind}(p_0)$, is defined in [5] as the winding number associated to the mapping $p \to \bar{r}(p)$. Near p_0 we may assume

$$f_1 = E_1, \; f_2 = \bar{E}_1, \; \tilde{f}_1 = qv_1, \; \tilde{f}_2 = \bar{r}v_1 .$$

It is then clear that $\tilde{f}(\Delta_2) = \epsilon(p_0)\operatorname{ind}(p_0)$ if $p_0 \in \Delta_2$, where $\epsilon(p_0) = \pm 1$, and

$$\sum \tilde{f}(\Delta_2) = \sum f(\Delta_2) + \sum_{p \in N} \epsilon(p)\operatorname{ind}(p) \quad .$$

Again, $\Sigma f(\Delta_2) = c_1(V^c)[N] = 0$ as $H^2(H,\mathbb{Z}) \cong \mathbb{Z}$. Thus, we have proved the following.

Proposition 2) *Let* V *be a real 2-plane subbundle of the complex 2-plane bundle* $\tilde{V}$ *over the compact, oriented surface* M. *If the set* N *is finite, then*

$$c_1(\tilde{V})[M] = \sum_{p \in N} \epsilon(p)\,\mathrm{ind}(p) , \tag{7}$$

where $\varepsilon(p) = \pm 1$.

We have deliberately left the sign $\epsilon(p)$ in (7) ambiguous in order to remain consistent with our previous definition [5] of $\mathrm{ind}(p)$, when $V = TM$ and $\tilde{V} = T\tilde{M}|_M$, for a real surface immersed in a complex 2-manifold $\tilde{M}$. In this case any choice of local orientation of M induces a natural local orientation of V. At a complex tangent $p \in M$ we choose the orientation to agree with the orientation of V_p as a complex line. We define $\mathrm{ind}(p)$ relative to this orientation. Then $\epsilon(p)$ is $+1$ or -1 according to whether this local orientation agrees or disagrees with the global orientation of M.

Now suppose that $\tilde{M}$ is a Kähler surface and that $M \subset \tilde{M}$ is an immersed minimal surface. If M is not a holomorphic curve, then the complex tangents are isolated and each has negative index [5]. If $\tilde{M} = \mathbb{P}_2\mathbb{C}$ and M is embedded of genus g and degree $k, M \sim k\mathbb{P}_1$ in $H_2(\mathbb{P}_2,\mathbb{Z})$, we have shown in [5] that

$$\sum_{p \in N} \mathrm{ind}(p) = k^2 + 2 - 2g . \tag{8}$$

We consider the case $g = 2$. By (8) $k = 0$ or $k = \pm 1$. Suppose $k = \pm 1$. Then by (8) there is a single complex tangent (at p) of index -1. Now $c_1(T\mathbb{P}_2) = 3c_1(L)$, where $L \to \mathbb{P}_2$ is the hyperplane section bundle, and $c_1(L)[M] = k$. By (7) $\epsilon(p)(-1) = \pm 3$, a contradiction. Thus $k = 0$, and (8) and (7) give

$$\text{ind}(p_1) + \text{ind}(p_2) = -2 \quad ,$$

$$\epsilon(p_1)\text{ind}(p_1) + \epsilon(p_2)\text{ind}(p_2) = 0 \quad .$$

It follows that $\text{ind}(p_1) = \text{ind}(p_2) = -1$, and $\epsilon(p_1) = -\epsilon(p_2)$. Since M can't be holomorphically embedded in $\mathbb{P}_2$, we have

Corollary 3) *Let M be a compact oriented surface of genus 2 minimally embedded in the complex projective plane. Then M has degree zero and precisely two complex tangents. Both have index -1 , one is positively oriented, and the other negatively oriented to M .*

I do not know whether such an embedding exists.

BIBLIOGRAPHY

1. S.S. Chern, "Characteristic classes of hermitian manifols", Ann. of Math., 47 (1946), 85-121.

2. J.W. Milnor and J.D. Stasheff, "Characteristic Classes", Ann. of Math. Studies No. 76, Princeton University Press, 1974.

3. N. Steenrod. The Topology of Fiber Bundles, Princeton Math. Series No. 14, Princeton University Press, 1951.

4. S.M. Webster, "The Euler and Pontrjagin numbers of an n-manifold in $\mathbb{C}^n$ ", Comment. Math. Helv. (to appear).

5. ___________, "Minimal surfaces in a Kähler surface", J. Differential Geometry (to appear).

SCHOOL OF MATHEMATICS
UNIVERSITY OF MINNESOTA
MINNEAPOLIS, MINNESOTA 55455

Contemporary Mathematics
Volume 49, 1986

HARMONIC MORPHISMS, FOLIATIONS AND GAUSS MAPS

J.C. Wood
School of Mathematics, University of Leeds, LS2 9JT, Great Britain

0. INTRODUCTION

A harmonic morphism is a map between Riemannian manifolds which preserves harmonic functions. The purpose of this paper is to show how to attach certain holomorphic objects to harmonic morphisms into a Riemann surface which are analogous to some known constructions for harmonic maps from a Riemann surface and to show how these objects can be used to study the harmonic morphisms.

The first object studied (section 2) is the horizontal quadratic differential analogous to the well-known quadratic differential for harmonic maps of a surface (see, for example, J.C. Wood [1]). This chapter grew out of ideas of P. Baird [2] and is mainly expository but with one application (Theorem 2.7). The higher dimensional analogue using stress energy is also studied and some examples of Baird are slightly clarified by Proposition 2.12.

Our main contributions are in the following sections (3,4,5) where we study the Gauss map (or section) of a harmonic morphism. We show (Corollary 3.8) that for a harmonic morphism ϕ from a four-dimensional Riemannian manifold M to a Riemann surface, the Gauss section has holomorphicity properties akin to those of a Gauss section (or Gauss lift) of a minimal surface in a four-dimensional manifold (generalizing the well-known anti-holomorphicity of the Gauss map of a minimal surface in Euclidean space

0271-4132/86 $1.00 + $.25 per page

(Chern [1]). These properties are most nicely expressed in terms of sections of the twistor bundles (Example 4.2) and give a bijective correspondence between equivalence classes of submersive harmonic morphisms and pairs of "± holomorphic" sections (Theorem 4.4) analogous to the main Theorem of Eells and Salamon [1]. These results really depend only on the properties of the foliation of the four-dimensional manifold M^4 defined by taking the fibres of ϕ; such foliations are conformal and minimal, the holomorphicity properties are given in Theorems 3.7 and 4.1.

In section 5, bearing in mind the well-known harmonicity of ± holomorphic maps between Kähler manifolds, we ask whether the Gauss section γ of a two-dimensional minimal foliation of a four-dimensional Riemannian manifold M^4 is harmonic. The answer is "not always", see Theorem 5.17. We also study the "fibre" tension field of the Gauss section of a foliation of arbitrary dimension and codimension, for example in Theorem 5.5 we characterize those Riemannian foliations with harmonic Gauss section. The method used is to calculate the trace of the second fundamental form of γ over various types of distributions (Propositions 5.1, 5.7 and 5.11). Note that if $M^4 \subset \mathbb{R}^4$ our results concern the usual tension field of the Gauss map.

Finally we draw attention to the work of Paul Baird [1,2,3] on harmonic morphisms, in particular, he uses the Gauss map to classify harmonic morphisms from an open subset of Euclidean 3-space to a surface. I am grateful to him for exchanging ideas on harmonic morphisms, I thank also J. Eells for helpful comments and Margaret Williams for typing my appalling handwriting.

1. HARMONIC MORPHISMS

Let (M^m, g) and (N^n, h) be smooth $(= C^\infty)$ Riemannian manifolds of dimensions m and n respectively. A continuous map $\phi : M \to N$ is called a harmonic morphism if, for each harmonic function $f : V \to \mathbb{R}$ defined on an open subset V of N with $\phi^{-1}(V)$ non-empty, $f \circ \phi : \phi^{-1}(V) \to \mathbb{R}$ is harmonic. Since, by Greene-Wu [1], we can choose smooth harmonic local coordinates on N any harmonic morphism must actually be smooth.

To characterize harmonic morphisms, a concept is required that generalizes that of Riemannian submersion. For any smooth map $\phi : M \to N$ let K denote the set of critical points of ϕ by which we mean the set of points where rank $d\phi < n$, points of $M \backslash K$ are called regular. For any $p \in M$, set $V_p = \ker d\phi_p$, V_p is called the vertical space at p; its orthogonal complement in T_pM, H_p, is called the horizontal space at p. A smooth map $\phi : M \to N$ is said to be horizontally (weakly) conformal (Ishihara [1]) (or semi-conformal (Fuglede [1])) if $d\phi = 0$ on K and $\phi|_{M\backslash K}$ is a conformal submersion (i.e. for all $p \in M\backslash K$, $d\phi_p : H_p \to T_{\phi(p)}N$ is surjective and $\|d\phi_p(X)\| = \lambda(p)\|X\| \; \forall X \in H_p$ for some $\lambda(p) \in (0,\infty)$. If we set $\lambda(p) = 0$ for $p \in K$ then $\lambda : M \to [0,\infty)$ is a continuous function called the conformality factor of the map ϕ, we see that $\lambda^2 = \frac{1}{n}\|d\phi\|^2$ is smooth. Note that (i) ϕ is a Riemannian submersion if and only if $\lambda \equiv 1$, (ii) if $m < n$ any horizontally weakly conformal mapping is constant, (iii) if $m = n$ a horizontally weakly conformal mapping is precisely a weakly conformal mapping, (iv) if $n = 1$ any smooth map is horizontally weakly conformal.

We now recall the characterization of harmonic morphisms.

Theorem 1.1 (Fuglede [1], Ishihara [1]). A smooth map $\phi : M \to N$ is a harmonic morphism if and only if it is (i) a harmonic map and (ii) horizontally weakly conformal.

In particular (i) if $\dim M < \dim N$ there are no non-constant harmonic morphisms from M to N, (ii) if $\dim M = \dim N = 2$, the harmonic morphisms are precisely the weakly conformal maps, (iii) if $\dim N = 1$ they are precisely the harmonic maps, (iv) if $\dim M = \dim N \geq 3$ they are precisely the homothetic maps (Fuglede, Ishihara) i.e. conformal maps with fixed conformality factor. Note also that unless a harmonic morphism $\phi : M \to N$ is constant, $M \setminus K$ is an open dense subset of M. For some special characterizations of harmonic morphisms into $\mathbb{R}^n$ or $\mathbb{C}$ see Fuglede [1], Bernard, Campbell and Davie [1]. For examples of harmonic morphisms and a classification of harmonic morphisms from U to $\mathbb{R}^2$ (or a Riemann surface) where U is an open subset of $\mathbb{R}^3$, see Baird [1,3].

Clearly the composition of harmonic morphisms is a harmonic morphism. In particular we have

Proposition 1.2 If $\phi : M \to N^2$ is a harmonic morphism to a two-dimensional Riemannian manifold and $\rho : N^2 \to P^2$ is a weakly conformal map between two-dimensional Riemannian manifolds then $\rho \circ \phi : M \to P^2$ is a harmonic morphism.

Thus, when discussing harmonic morphisms to a two-dimensional Riemannian manifold (N^2, h), we need only use the conformal structure on N^2 induced by h, in particular the concept of harmonic morphism to a Riemann surface is well-defined (c.f. harmonic maps from a Riemann surface - J.C. Wood [1]).

To construct explicit examples of harmonic morphisms we recall a result of Baird and Eells [1] that a horizontally conformal submersion $\phi : M \to N$

with minimal fibres and <u>either</u> dim $N = 2$ <u>or</u> grad λ^2 vertical is a harmonic morphism (see also §2 below). This shows that the following are harmonic morphisms : (i) the radial projections $\mathbb{R}^m \setminus \{0\} \to S^{m-1}$ $(m = 1,2,\ldots)$ defined by $x \to x/|x|$; (ii) the Hopf maps $S^3 \to S^2$ and $S^{2n+1} \to CP^n$ $(n = 1,2,\ldots)$; (iii) any holomorphic map from a Kähler manifold to a Riemann surface. Note that in the first two cases the fibres are totally geodesic and in the last case they are complex hypersurfaces. For more harmonic morphisms see examples 2.14 and 3.9 below.

<u>Notation</u>. $M = M^m$ and $N = N^n$ will always denote smooth $(=C^\infty)$ manifolds of dimension m and n respectively. Given a smooth vector bundle E over M, $C^\infty(E)$ will denote the space of smooth (local or global) sections of E, $\langle\ ,\ \rangle$ and ∇(or ∇^E) will denote a fibre metric and connection on E, and for $v \in E$ we write $\|v\| = \sqrt{\langle v,v\rangle}$. On a Riemannian manifold $M = (M,g)$ with metric tensor g, we write $\langle X,Y\rangle = g(X,Y)$ for $X,Y \in T_pM, p \in M$ and ∇^M will denote the Levi-Civita connection on (M,g). Given a vector field X and a tensor field Z on M, L_XZ will denote the Lie derivative of Z in direction X. In computations involving bases, the double summation convention will frequently be used without comment.

For a map $\phi : M^m \to N^n$ with $m \geq n$ K will denote the set of singular points of ϕ by which we mean points $p \in M$ where $d\phi_p$ fails to be surjective.

For background information on harmonic maps, harmonic morphisms and foliations we recommend Eells-Lemaire [1,2], Baird [1] and Reinhart [1] respectively.

2. THE HORIZONTAL QUADRATIC DIFFERENTIAL In this chapter we develop ideas of Baird [2].

We assume from now on that $m(=\dim M) \geq n(=\dim N)$. Let $\phi : M^m \to N^n$ be a smooth map with rank n somewhere. Say that ϕ has minimal fibres if at each regular point p the fibre through p, $\phi^{-1}(\phi(p))$ has zero mean curvature. (Note that the fibre is a submanifold of dimension m-n on a neighbourhood of p. Baird-Eells show, using stress energy:

Proposition 2.1 (Baird-Eells [1]) Let $\phi : M^m \to N^2$ be a non-constant horizontally weakly conformal mapping to a Riemann surface. Then ϕ is harmonic if and only if it has minimal fibres.

We now show that the result ϕ harmonic <=> ϕ has minimal fibres is valid for a larger class of mappings ϕ namely those with holomorphic horizontal quadratic differential.

To proceed, we assume that N^2 is oriented, (if N^2 is non-orientable we can proceed locally - results will be independent of the orientation chosen). Let $\phi : M^m \to N^2$ be a smooth map with rank 2 somewhere. At each regular point $p \in M$ orient the horizontal space H_p by demanding that $d\phi_p : H_p \to T_{\phi(p)}N$ be orientation preserving. Now define an almost complex structure $J^H = J^H_p$ on H_p by setting J^H_p equal to rotation through $+\frac{\pi}{2}$. Writing $H^c_p = H_p \otimes \mathbb{C}$ and extending J^H to H^c_p by complex linearity, set H'_p (resp. H''_p) equal to the $+i$ (resp. $-i$) eigenspace of $J^H_p : H^c_p \to H^c_p$. Thus $H'_p = \{Z = X - iJ^H_p X : X \in H_p\}$. Clearly H'_p is the fibre at p of a vector bundle H' over $M\backslash K$ (and similarly for H_p, H^c_p and H''_p). Now extend the differential $d\phi_p$ to a complex linear map $d^c\phi_p : H^c_p \to T^c_{\phi(p)}N$ ($= T_{\phi(p)}N \otimes C$) and define a section η of the bundle $\mathcal{L}^2(H',\mathbb{C})$ over $M\backslash K$ by (*)

(*) As always, K will denote the set of points p where rank $d\phi(p) < 2$.

$$\eta(Z,Z') = \langle d^c\phi(Z), d^c\phi(Z')\rangle .$$

We shall call η the <u>horizontal quadratic differential of</u> ϕ. Note that η is symmetric : $\eta(Z',Z) = \eta(Z,Z')$ for all $Z,Z' \in H'_p$ and thus η is actually a section of the symmetric square $\odot^2(H')^*$. We have

<u>Proposition 2.2</u> Let $\phi : M^m \to N^2$ be a smooth map with $M\backslash K$ dense in M. Then the horizontal quadratic differential $\eta \equiv 0$ on $M\backslash K$ if and only if ϕ is horizontally weakly conformal on M. (Note that if ϕ is a real-analytic map with rank 2 somewhere then $M\backslash K$ is always dense.)

<u>Proof</u> Let $p \in M\backslash K$. Choose an oriented orthonormal basis X, Y for the horizontal space H_p and write $Z = X - iY$ so that $Z \in H'_p$. Then a simple calculation yields

$$\eta(Z,Z) = \langle d\phi(X), d\phi(X)\rangle - \langle d\phi(Y), d\phi(Y)\rangle - 2i\langle d\phi(X), d\phi(Y)\rangle .$$

The right-hand side is zero if and only if $d\phi : H_p \to T_{\phi(p)}N$ is conformal; the proposition follows.

Let us say that η is <u>horizontally holomorphic</u> if $\nabla_{\bar{Z}}\eta = 0$ for all $p \in M\backslash K$, $Z \in H'_p$.

<u>Proposition 2.3</u> The horizontal quadratic differential η is horizontally holomorphic on $M\backslash K$ if and only if

$$\mathrm{Trace}\ \nabla d\phi|_{H_p \times H_p} = 0$$

for all $p \in M\backslash K$.

<u>Proof</u> Let $p \in M\backslash K$. As above write $Z = X - iY$. Extend Z to a smooth section of H' in some neighbourhood of p (still denoted by Z). Then denoting by $\nabla^{H'}_Z$ the connection on H' obtained from the Levi-Civita connection ∇^M by projection viz : $\nabla^{H'}_Z X = \{\mathcal{H}(\nabla^M_Z X)\}^{(1,0)}$ for $Z \in C^\infty(TM)$, $X \in C^\infty(H')$ where $\mathcal{H} : T^cM \to H^c$ is the complex linear extension of orthogonal projection $TM \to H$ and $^{(1,0)}$ is the projection $H^c \to H'$ along H''.

$(\nabla_{\bar Z}\eta)(Z,Z)$

$$= \nabla_{\bar Z}\{\eta(Z,Z)\} - 2\eta(\nabla^{H'}_{\bar Z} Z,Z)$$

$$= \nabla_{\bar Z}\{\eta(Z,Z)\} - 2\eta(\nabla^{M}_{\bar Z}Z,Z)$$

$$= \nabla_{\bar Z}\langle d^c\phi(Z),d^c\phi(Z)\rangle - 2\langle d^c\phi(\nabla^{M}_{\bar Z}Z),d^c\phi(Z)\rangle$$

$$= 2\langle\nabla^{N}_{\bar Z}d^c\phi(Z) - d^c\phi(\nabla^{M}_{\bar Z}Z),d^c\phi(Z)\rangle$$

$$= 2\langle(\nabla^{N}_{\bar Z}d^c\phi)(Z),d^c\phi(Z)\rangle$$

$$= 2\langle\nabla d\phi(X,X) + \nabla d\phi(Y,Y),d\phi(X) - id\phi(Y)\rangle$$

$$= 2\langle\text{Trace } \nabla d\phi|_{H_p\times H_p},\ d\phi(X) - id\phi(Y)\rangle$$

Since $d\phi(X)$ and $d\phi(Y)$ together span $T_{\phi(p)}N$ the inner product is zero if and only if $\text{Trace } \nabla d\phi|_{H_p\times H_p} = 0$. The Proposition follows.

Remark 2.4 (i) To get from the first to the second line we need to know that if Z is a section of H' then so is $\nabla^{H'}_{\bar Z}Z$. This follows easily from the useful

Lemma 2.5 Let (E,∇) be a Riemannian connected vector bundle with oriented fibres of dimension 2 over a smooth manifold M. For each $p \in M$ let $J : E_p \to E_p$ be rotation through $+\frac{\pi}{2}$ and let E' (resp. E'') be the $+i$ (resp. $-i$) eigenspaces of the complex linear extension of J to $E^c_p = E_p \otimes \mathbb{C}$. Then J is parallel i.e. $(\nabla_X J)(Y) \equiv \nabla_X(J(Y)) - J(\nabla_X Y) = 0$ for all $X \in C^\infty(TM)$ $Y \in C^\infty(E)$. Consequently so are E' and E'', i.e. if $X \in C^\infty(TM)$ (or $C^\infty(T^cM)$) and $Y \in C^\infty(E')$ then $\nabla_X Y \in C^\infty(E')$ and similarly for E''.

Proof Elementary.

Note also that, if we define $\tilde{\eta}$ to be a section of $\mathcal{L}^2(T^cM,C)$ by $\tilde{\eta}(Z,\bar{Z}) = \eta(\mathcal{H}(Z)^{1,0}, \mathcal{H}(\bar{Z})^{1,0})$ where $\mathcal{H}: T^cM \to H^c$ is the complex linear extension of orthogonal projection $TM \to H$ of a vector onto the horizontal subspace and ^{1,0} denotes that we then take the part in H', then $\tilde{\eta}$ is defined on the whole of M and Proposition 2.2 holds in the form $\tilde{\eta} \equiv 0$ on $M \Leftrightarrow \phi$ horizontally weakly conformal. However Proposition 2.3 does not have a nice formulation in terms of $\tilde{\eta}$.

(ii) The last two Propositions are analogous to results on the holomorphic quadratic differential for maps $M^2 \to N^n$, see, for example, J.C. Wood [1].

Theorem 2.6 Let $\phi : M^m \to N^2$ be a smooth map with rank 2 somewhere. Then any two of the following conditions imply the other one.

(i) ϕ is harmonic on M,

(ii) ϕ has minimal fibres,

(iii) η is horizontally holomorphic on $M\backslash K$.

Proof At any $p \in M\backslash K$, the tension field of ϕ,

$\tau(\phi) = \text{Trace } \nabla d\phi = \text{Trace } \nabla d\phi|_{H_p \times H_p} + \text{Trace } \nabla d\phi|_{V_p \times V_p}$.

Let $e_3, \ldots, e_m$ be an orthonormal frame for V in a neighbourhood of p. Then

$$\begin{aligned}\text{Trace } \nabla d\phi|_{V_p \times V_p} &= \sum_{\alpha=3}^{m} \nabla^N_{e_\alpha} d\phi(e_\alpha) - d\phi(\nabla^M_{e_\alpha} e_\alpha) \\ &= 0 - d\phi(\text{mean curvature of the fibre through } p).\end{aligned}$$

Noting that the mean curvature of the fibre is horizontal, it follows that

Trace $\nabla d\phi|_{V_p \times V_p} = 0$ if and only if the fibre is minimal at p.

Finally note that if ϕ is harmonic on $M\backslash K$ then K can contain no open set otherwise by unique continuation (Sampson [1]) ϕ would have rank ≤ 1 everywhere. It follows that $M\backslash K$ is dense so that ϕ is harmonic on M.

As an application:

Theorem 2.7 Let $\phi : M^m \to (S^2,h)$ be a harmonic submersion to the 2-sphere with any Riemannian metric. Suppose ϕ has minimal fibres and integrable horizontal distribution with leaves homeomorphic to 2-spheres. Then ϕ must be a harmonic morphism.

Proof By Proposition 2.3 the horizontal quadratic differential of ϕ gives a holomorphic differential on each leaf of the horizontal distribution. But, as is well known, a holomorphic quadratic differential on a 2-sphere must vanish, thus by Proposition 2.2, ϕ is horizontally conformal and thus a harmonic morphism.

We give a generalization to higher dimensions which also grew out of work of Baird [2]. Let $\phi : (M^m,g) \to (N^n,h)$ be a smooth map between Riemannian manifolds. The stress energy tensor S_ϕ is the 2-tensor on M defined by $S_\phi = e_\phi g - \phi^*h$ where $e_\phi = \frac{1}{2}\|d\phi\|^2$ denotes the energy density of ϕ and ϕ^*h denotes the pull back of the tensor h to a tensor on M, see Baird-Eells [1] for more details.

Lemma 2.8 Let $\phi : M^m \to N^n$ be a smooth map of rank n somewhere. Then for any regular point $p \in M^m$ and $Y \in H_p$,

$$\sum_{i=1}^{n} (\nabla_{e_i} S_\phi)(e_i, Y) = - \langle \mathrm{Tr}\, \nabla d\phi|_{H_p \times H_p}, d\phi(Y)\rangle.$$

Here $\{e_i\}$ is any orthonormal basis for the horizontal space H_p and $Y \in H_p$.

This motivates the definition: say S_ϕ is <u>horizontally divergence free</u> if $\sum_{i=1}^{n} (\nabla_{e_i} S_\phi)(e_i, Y) = 0 \quad \forall\, Y \in H_p$.

<u>Theorem 2.9</u> Let $\phi : M^m \to N^n$ be a smooth map of rank n somewhere with dense set of regular points $M \backslash K$. Then any two of the following conditions imply the other one:

(i) ϕ is harmonic on M,

(ii) ϕ has minimal fibres,

(iii) S_ϕ is horizontally divergence free on $M \backslash K$.

<u>Proof</u> Similar to Theorem 2.6.

<u>Remarks 2.10</u> (i) If ϕ is real analytic the set of regular points will automatically be dense.

(ii) The properties (i), (ii) and (iii) in Theorems 2.9 (resp. 2.6) are unchanged if ϕ is composed with a totally geodesic map $N^2 \to P^2$ (resp. $N^n \to P^p$). For example if $\phi : M^m \to \mathbb{R}^2$ is a horizontally weakly conformal, then composing with a linear map $\mathbb{R}^2 \to \mathbb{R}^2$ destroys this property but does not alter the holomorphicity of the horizontal quadratic differential.

The following result is essentially a restatement of a result of Baird-Eells [1]:

<u>Proposition 2.11</u> Suppose that ϕ is horizontally weakly conformal. Then S_ϕ is horizontally divergence free if and only if $\dim N = 2$ or $\operatorname{grad} \lambda^2$ is vertical.

<u>Proof</u> Let $p \in M \backslash K$ and let $\{e_i\}$ be an orthonormal basis for H_p. Then omitting summation signs over $i = 1, \ldots, n$:

Then $(\nabla_{e_i} S_\phi)(e_i, e_j) = (\nabla_{e_i}(e_\phi g))(e_i, e_j) - (\nabla_{e_i} \phi^* h)(e_i, e_j)$.

But $\nabla_{e_i} g = 0$ for all i so that $\nabla_{e_i}(e_\phi g)(e_i, e_j) = \nabla_{e_j} e_\phi$

$$(\nabla_{e_i} \phi^* h)(e_i, e_j) = \nabla_{e_i}(\phi^* h(e_i, e_j)) - \phi^* h(\nabla^M_{e_i} e_i, e_j) - \phi^* h(e_i, \nabla^M_{e_i} e_j)$$

$$= \nabla_{e_j} \lambda^2 - \{<\nabla^M_{e_i} e_i, e_j> + <e_i, \nabla^M_{e_i} e_j\} \lambda^2$$

$$= \nabla_{e_j} \lambda^2 .$$

Thus $(\nabla_{e_i} S_\phi)(e_i, e_j) = \nabla_{e_j} e_\phi - \nabla_{e_j} \lambda^2 = (\frac{m}{2} - 1) \nabla_{e_j} \lambda^2$.

The Proposition follows.

We thus recover results of Baird-Eells [1] and Baird [1,2] that *if ϕ is a horizontally conformal submersion any two of the following conditions imply the other*:

(i) ϕ *harmonic*, (ii) ϕ *has minimal fibres*, (iii) *grad λ^2 is vertical*.

For a generalization, following Reinhart [1] call a subbundle E of TM *totally geodesic* if $\nabla^M_X X \in C^\infty(E)$ for all $X \in C^\infty(E)$, (see also §3), more generally if we have a further subbundle E' of E say that E' is *totally geodesic in* E if $\nabla^E_X X \in C^\infty(E')$ for all $X \in C^\infty(E')$, (here ∇^E is the induced connection on E obtained from ∇^M by projection.

Proposition 2.12 Suppose that $\phi : M^m \to N^n$ is a smooth submersion such that the horizontal bundle H has a decomposition $H = \sum_{\mu=1}^{\nu} H_\mu$ into subbundles which are totally geodesic in H. Suppose further that ϕ is horizontally homothetic on each H_μ i.e. there are functions $\lambda_\mu : M \to (0,\infty)$ with grad λ_μ vertical such that

$$\|d\phi(X)\| = \lambda_\mu(p)\|X\| \qquad \forall\, p \in M,\ X \in (H_\mu)_p .$$

Then S_ϕ is horizontally divergence free.

Hence ϕ is harmonic if and only if it has minimal fibres.

Proof Proceeding as in Proposition 2.11,

$$\nabla_{e_i}(e_\phi g)(e_i,e_j) = \nabla_{e_j} e_\phi \quad \text{as before and}$$

$$(\nabla_{e_i}\phi^* h)(e_i,e_j) = \nabla_{e_j}\lambda^2_{\mu_j} - \langle\nabla^M_{e_i} e_i,e_j\rangle(\lambda^2_{\mu_i} - \lambda^2_{\mu_j}).$$

If $\mu_i = \mu_j$ the last term is clearly zero, if $\mu_i \neq \mu_j$ $\langle\nabla^M_{e_i} e_i,e_j\rangle = \langle(\nabla^M_{e_i} e_i),e_j\rangle = 0$ by total geodesity of H_{μ_j}, hence,

$$(\nabla_{e_i} S_\phi)(e_i,e_j) = \nabla_{e_j}(e_\phi - \lambda^2_{\mu_j}) = 0.$$

Remark 2.13 Compare with Baird [1] Proposition 2.2.5 and Theorem 7.3.1.

Example 2.14 (Baird [1]) Let S^m be equipped with an isoparametric system of hypersurfaces, let V_1, V_2 denote the focal varieties (if only one let $V_2 = \phi$) and let $\Pi : M\backslash V_2 \to V_1$ denote projection down the integral curves. Then the decomposition of the horizontal spaces into eigenspaces of the principal curvatures satisfies the above conditions, hence Π is horizontally divergence free. But since the fibres are minimal (in fact totally geodesic) Π is harmonic by Proposition 2.9. If there are no more than two distinct principal curvatures Π is horizontally conformal and so is a harmonic morphism.

For example for any integers $p,q,n \geq 1$ with $p + q = n$ write each point of S^{n-1} as $((\cos t)x,(\sin t)y) \in \mathbb{R}^p \times \mathbb{R}^q = \mathbb{R}^n$ where $x \in S^{p-1}$, $y \in S^{q-1}$, $t \in [0,\frac{\pi}{2}]$; this expresses S^{n-1} as the join of S^{p-1} and S^{q-1}. Then t is an isoparametric function on S^{n-1} with focal varieties given by $S^{p-1} \times \{0\}$ and $\{0\} \times S^{q-1}$ and level sets $(\cos t)\, S^{p-1} \times (\sin t)\, S^{q-1}$ having at most two principal curvatures. The map $\Pi : S^{n-1} \setminus S^{p-1} \times \{0\} \to \{0\} \times S^{q-1}$ defined by $((\cos t)x,(\sin t)) \to y$ is thus a harmonic morphism.

3. GAUSS MAPS OF FOLIATIONS AND HARMONIC MORPHISMS

Let M^m be a smooth Riemannian manifold. By a (smooth) distribution V of dimension k (and codimension $m-k$) we mean a (smooth) subbundle of rank k of the tangent bundle. If the distribution is integrable, i.e. $[X,Y] \in C^\infty(V)$ $\forall$ $X,Y \in C^\infty(V)$, it is called a foliation and the connected components of the integral submanifolds of V are called the leaves of the foliation.

Example 3.1 Given any smooth submersion $\phi : M^m \to N^n$, for each $p \in M$, set V_p = vertical space at $p = \ker (d\phi_p)$. Then $V = \{V_p\}$ is an $(m-n)$-dimensional smooth distribution which is integrable with leaves equal to the components of the fibres $\phi^{-1}(q)$, $q \in N$ of ϕ. We call V the foliation associated to ϕ.

Let V be a smooth distribution on M^m of dimension k. Then setting $H_p = V_p^{\perp}$ for each $p \in M$ defines a smooth distribution of dimension $m-k$; borrowing terminology from example 3.1 we shall call the original distribution V the vertical distribution and the distribution H associated to it the horizontal distribution; for each $p \in M$, V_p and H_p will be called the vertical and horizontal spaces at p, respectively. Note that integrability of V does not imply integrability of H.

Let $G_k(TM)$ denote the Grassman bundle over M with fibre at p equal to the set $G_k(T_pM)$ of all k-dimensional subspaces of the tangent space T_pM. Associated to a smooth distribution V of dimension k on M^m we have the Gauss section $\gamma : M \to G_k(TM)$, given by $\gamma(p) = V_p$, the Gauss section of the associated horizontal distribution H will be denoted by $\gamma^{\perp} : M \to G_{m-k}(TM)$ $\gamma^{\perp}(p) = H_p$. Note that if M is an open subset of a Euclidean space the

Grassman bundle is trivial and we obtain the usual Gauss map of M, $M \to G_k(\mathbb{R}^m)$ from the Gauss section by projecting $G_k(TM)$ onto $G_k(\mathbb{R}^m)$ using the canonical isomorphism $G_k(TM) = M \times G_k(\mathbb{R}^m)$. The bundle $G_k(TM)$ inherits a connection from the Levi-Civita connection on M, for any $p \in M$, $V_p \in G_k(T_pM)$, this defines a projection $^f : T_{V_p}G_k(TM) \to T^f_{V_p}G_k(TM)$ of the full tangent space at V_p onto the <u>tangent space to the fibre</u> $T^f_{V_p}G_k(TM) = T_{V_p}(G_k(T_pM))$, we may thus consider the fibre component of the differential of γ at a point $p \in M$:

$$d\gamma^f : T_pM \xrightarrow{d\gamma} T_{V_p}G_k(TM) \xrightarrow{f} T^f_{V_p}G_k(TM).$$

Now for each $p \in M$ and $V_p \in G_k(T_pM)$, the tangent space to the fibre $T^f_{V_p}G_k(TM)$ may be identified with the vector space $\mathcal{L}(V_p, H_p)$ of linear maps from V_p to $H_p = V_p^{\perp}$. Under this identification $d\gamma^f$ is given by

$$d\gamma^f(X) \in \mathcal{L}(V_p, H_p) \qquad (X \in T_pM)$$

$$d\gamma^f(X)(W) = \mathcal{H}(\nabla^M_X W) \qquad (W \in V_p)$$

where W is extended to a local section of V to calculate the right-hand side, the result being independent of the extension chosen. See C.M. Wood's thesis [1] for more details of this identification which we shall use frequently without comment.

We now study the holomorphicity properties of the Gauss section of a 2-dimensional distribution V. First orient V i.e choose a consistent orientation on each space V_p. This is always possible locally, and as our results will not depend on what orientation is chosen, we can, without loss of generality, assume that we can choose an orientation globally. Now for each $p \in M$ define an almost complex structure J^V on each space V_p by rotation through $+\frac{\pi}{2}$. Give each space $T^f_{V_p}G_2(TM) = \mathcal{L}(V_p, H_p)$ a corresponding almost complex structure J^f by

$$(1) \qquad J^f(A) = -A \circ J^V, \qquad (A \in \mathcal{L}(V_p, H_p)).$$

(The minus sign is to accord with historical conventions but is not essential.) Say that a section $\gamma : M \to G_2(TM)$ is (a) vertically holomorphic (resp. antiholomorphic) (section) if

$$d\gamma^f(J^V W) = J^f(d\gamma^f(W)) \qquad \text{(resp. minus this)}$$

for all $p \in M$, $W \in V_p$. Note that vertically refers to the restriction $W \in$ vertical distribution rather than to the fact that we are considering the fibre component $d\gamma^f$ of $d\gamma$. Note also that vertical holomorphicity is independent of the actual orientation chosen on each V_p since changing the orientation changes the signs of J^V and J^f. Similarly for vertical antiholomorphicity. If γ is vertically holomorphic and antiholomorphic then it is vertically constant i.e. $d\gamma^f(W) = 0$ for all $p \in M$, $W \in V_p$.

We now interpret vertical holomorphicity and antiholomorphicity. To do this we recall some definitions which we may make for a distribution V of arbitrary dimension, see, for example, Reinhart [1].

By the second fundamental form of V at $p \in M$ we mean the symmetric bilinear form

$$\zeta = \zeta_p : V_p \times V_p \longrightarrow H_p$$

$$(X,Y) \longmapsto \tfrac{1}{2}\mathcal{H}(\overset{M}{\nabla}_X Y + \overset{M}{\nabla}_Y X).$$

To calculate the right-hand side, X and Y must be extended to local sections of V, the result is independent of the chosen extensions. Note that if V is integrable $\zeta_p(X,Y) = \mathcal{H}(\overset{M}{\nabla}_X Y)$ the usual second fundamental form of the leaves. Whether V is integrable or not $\zeta_p(X,X)$ measures the normal component of the acceleration vector of any smooth curve $\gamma : (-\varepsilon,\varepsilon) \to M$ with $\gamma(0) = p$, $\gamma'(0) = X$ and $\gamma'(t)$ lying in V for all $t \in (-\varepsilon,\varepsilon)$, if $\|X\| = 1$, $\zeta_p(X,X)$ = normal component of the curvature of γ and is called the <u>normal curvature of</u> V <u>at</u> p <u>in direction</u> X. The vector $\frac{1}{k}$ Trace $\zeta_p = \frac{1}{k} \sum_{e_\alpha} \mathcal{H}(\nabla^M_{e_\alpha} e_\alpha)$ where $\{e_\alpha\}$ is an orthonormal basis and $k = \dim V$ is called <u>the mean curvature of</u> V <u>at</u> p.

<u>Definitions 3.2</u> A distribution V is said to be (i) <u>minimal</u> if its mean curvature vanishes;

(ii) <u>umbilic</u> if, for each $p \in M$, the normal curvature in direction X, $\zeta_p(X,X)$, is independent of $X \in T_pM$, $\|X\| = 1$;

(iii) <u>totally geodesic</u> if, for each $p \in M$, $\zeta_p \equiv 0$, equivalently, all normal curvatures are zero.

Note that a distribution is totally geodesic if and only if any geodesic of M which is tangent to the distribution at some point remains tangent to it for its entire length. If the distribution is integrable then it is totally geodesic if and only if its leaves are totally geodesic in the usual sense.

<u>Proposition 3.3</u> Let V be a 2-dimensional distribution on a smooth Riemannian manifold of arbitrary dimension. Then the Gauss section $\gamma : M \to G_2(TM)$ is (a) vertically holomorphic, (b) vertically anti-holomorphic, (c) vertically constant, according as the distribution V is (a) umbilic, (b) integrable and minimal, (c) integrable and totally geodesic.

<u>Proof</u> Let $p \in M$. Let $X, Y = J^V X$ be an oriented orthonormal frame for V in some neighbourhood of p. Then γ is vertically holomorphic at p

if and only if

$$d\gamma^f(J^V W)(W') = - d\gamma^f(W)(J^V W')$$

for any $W, W' \in V_p$. In terms of the frame X, Y this is equivalent to

$$d\gamma^f(X)X = d\gamma^f(Y)Y$$

and $$d\gamma^f(X)Y = - d\gamma^f(Y)X$$

i.e. $$\mathcal{H}(\nabla^M_X X - \nabla^M_Y Y) = 0$$

and $$\mathcal{H}(\nabla^M_X Y + \nabla^M_Y X) = 0$$

which is easily seen to be the condition of umbilicity at p.

Similarly γ is vertically antiholomorphic at p if and only if

$$\mathcal{H}(\nabla^M_X X + \nabla^M_Y Y) = 0$$

and $$\mathcal{H}(\nabla^M_X Y - \nabla^M_Y X) = 0.$$

The first of these conditions is minimality at p, the second is integrability.

Lastly γ is vertically constant if and only if $\mathcal{H}(\nabla^M_W W') = 0$ for all sections W, W' of V. This is equivalent to $\zeta(W, W') = 0$ and $[W, W'] = 0$ i.e. total geodesity and integrability.

Remark 3.4 If $M = \mathbb{R}^m$ the Gauss section may be replaced by the Gauss map. The antiholomorphicity of the Gauss map of a surface in Euclidean space is well-known (see for example Chern [1]) (and determined our choice of minus sign in the definition of J^f). For related results on $\pm$ holomorphicity and umbilicity see Eells-Salamon[1,2].

The interesting development comes when we consider the foliations which arise from a harmonic morphism $\phi : M^4 \to N^2$. First some terminology.

Let V be a distribution of arbitrary dimension in a smooth manifold M and let H denote the associated horizontal distribution. Say that the distribution V is (a) <u>Riemannian</u>, (resp. (b)<u>conformal</u>) according as

$$\text{(a)} \quad (L_W g)(X,Y) = 0$$

$$\text{(resp. (b)} \quad (L_W g)(X,Y) = \lambda(W) g(X,Y))$$

for all $p \in M$, $X,Y \in H_p$, $W \in V_p$ where $\lambda(W)$ is a non-negative number which depends only on W and not on X and Y. Note that to calculate $(L_W g)(X,Y)$ we must extend W, X and Y to sections of V, H respectively (see the proof of Proposition 3.5) but $(L_W g)(X,Y)$ does not depend on the choice of extensions.

Note also that the <u>Bott partial connection on</u> H is defined, by $\overset{\circ}{\nabla}_W X = \mathcal{H}(L_W X)$, $W \in C^\infty(V)$, $X \in C^\infty(H)$. Then, in terms of the induced partial connection on $\odot^2 H^*$ (b) reads

$$\overset{\circ}{\nabla}_W g = \lambda(W) g .$$

We may relate properties of a distribution V to its associated horizontal distribution:

<u>Proposition 3.5</u> A distribution V is (a) Riemannian, (b) conformal according as the associated horizontal distribution H is (a) totally geodesic, (b) umbilic. Further, if H is integrable the distribution V is Riemannian if and only if the vertical spaces are parallel in the horizontal direction i.e. $\nabla_X W \in C^\infty(V)$ $\forall$ $X \in C^\infty(H)$, $W \in C^\infty(V)$.

<u>Proof</u> Let $p \in M$, $X,Y \in H_p$, $W \in V_p$. Extend X and Y (resp. W) to local sections of H (resp. V). Then

$$
\begin{aligned}
(L_W g)(X,Y) &= \nabla_W \langle X,Y\rangle - \langle L_W X, Y\rangle - \langle X, L_W Y\rangle \\
&= \nabla_W \langle X,Y\rangle - \langle \nabla^M_W X, Y\rangle - \langle X, \nabla^M_W Y\rangle \\
&\qquad - \langle \nabla^M_X Y\rangle - \langle X, \nabla^M_X Y\rangle \\
&= 0 \quad - \langle \nabla^M_X Y + \nabla^M_Y X, W\rangle
\end{aligned}
$$

using the fact that ∇^M is Riemannian

$$(L_W g)(X,Y) = -\langle \zeta(X,Y), W\rangle .$$

Thus $(L_W g)(X,Y) = \lambda(W) g(X,Y)$ if and only if

$$\langle \zeta(X,Y), W\rangle = -\lambda(W)\langle X,Y\rangle \qquad \forall\ X,Y \in H_p,\quad W \in V_p .$$

This holds if and only if

$$\langle \zeta(X,X), W\rangle = -\lambda(W) \qquad \forall\ X \in H_p,\ \ \|X\| = 1,\ \ W \in V_p$$

which is clearly equivalent to umbilicity if $\lambda(W) \neq 0$ for some $W \in V_p$ and total geodesity if $\lambda(W) = 0$ for all $W \in V_p$.

For the last assertion merely note that

$$\langle \nabla^M_X W, Y\rangle = -\langle W, \nabla^M_X Y\rangle$$

for all $X,Y \in C^\infty(H)$, $W \in C^\infty(V)$ and, with H integrable, the right-hand side is zero for all X,Y,W if and only if H is totally geodesic.

<u>Remark 3.6</u> Result (a) above says that H has totally geodesic leaves if and only if $L_W(g|_H) = 0$ for all $p \in M$, $W \in V_p$. As a useful generalization for constructing examples note that H <u>has minimal leaves if and only if</u> $L_W(\nu|_H) = 0$ <u>for all</u> $p \in M$, $W \in V_p$. Here $\nu|_H$ denotes the volume form on M restricted to H. This generalization follows from the equalities

$$
\begin{aligned}
L_W(\omega|_H)(e_1,\dots,e_m) &= \sum_{i=1}^m \langle L_W e_i, e_i\rangle \\
&= \langle \sum_{i=1}^m \nabla_{e_i} e_i, W\rangle
\end{aligned}
$$

where $e_1,\dots,e_m$ is an orthonormal frame for H on a neighbourhood of p and W is extended to a local section of V.

Now let us consider a 2-dimensional distribution V in a 4-dimensional Riemannian manifold M^4 and let H be the associated 2-dimensional horizontal distribution. As we did for V we may now give each space H_p, $p \in M$ an almost complex structure J^H and then we say that the Gauss section of H, $\gamma^\perp : M \to G_2(TM^4)$, is <u>horizontally holomorphic</u> if $d\gamma^{\perp f}(J^H X) = J^f d\gamma^{\perp f}(X)$ $\forall\, p \in M,\ X \in H_p$.

<u>Theorem 3.7</u> Let V be a 2-dimensional distribution in a 4-dimensional Riemannian manifold M^4. Then

(i) V is integrable and minimal if and only if its Gauss section $\gamma : M \to G_2(TM^4)$ is vertically antiholomorphic.

(ii) V is conformal if and only if the Gauss section $\gamma^\perp : M \to G_2(TM^4)$ is horizontally holomorphic.

Note that $\gamma^\perp$ being horizontally holomorphic is equivalent to γ being horizontally holomorphic but with respect to a new almost complex structure $\tilde{J}^f$ on $T^f G_2(TM)$ namely, under the identification $T^f_{V_p} G_2(TM) = \mathcal{L}(V_p, H_p)$ we set

$$\tilde{J}^f(A) = J^H \circ A \qquad \forall\, A \in \mathcal{L}(V_p, H_p). \tag{2}$$

Note that with the almost complex structures $J^f, \tilde{J}^f$ defined above the map $V \to V^\perp$ from $(G_2(TM), J^f)$ to $(G_2(TM), \tilde{J}^f)$ is holomorphic on each fibre.

Finally we consider the Gauss section of a harmonic morphism. By the Gauss sections $\gamma : M^m \to G_{m-n}(TM)$ and $\gamma^\perp : M^m \to G_m(TM)$ of a submersion $\phi : M^m \to N^n$ we mean the Gauss sections of the associated vertical and horizontal distributions (see example 3.1).

<u>Corollary 3.8</u> Let $\phi : M^4 \to N^2$ be a submersion which is a harmonic morphism.

Then $\gamma : M^4 \to G_2(TM^4)$ is vertically antiholomorphic and $\gamma^\perp : M^4 \to G_2(TM^4)$ is horizontally holomorphic.

Examples 3.9 Examples of submersions which are harmonic morphisms $M^4 \to N^2$:

(i) Let (M^4,g,J) be a 4 (real) dimensional Kähler manifold and let $\phi : M^4 \to N^2$ be a holomorphic submersion to a Riemann surface. Then ϕ is a harmonic morphism. Note that if the resulting distributions V,H of vertical and horizontal spaces are correctly oriented, the almost complex structures J^V and J^H discussed above agree with J. The Gauss map of such a ϕ will be discussed in §4 (example 4.4).

(ii) Given any Riemannian manifolds (N,g_N) and (P,g_P) set $M = N \times P$ with product metric $g_M = (g_N,g_P)$ so that $\phi : (M,g_M) \to (N,g_N)$ is a Riemannian submersion with totally geodesic fibres (and totally geodesic integrable horizontal spaces). Now given any Riemannian submersion $\phi : (M,g_M) \to (N,g_N)$ with totally geodesic fibres we may change the metric of M^4 on each vertical space V_p in an arbitrary (smooth) fashion. Such a change will not affect the Riemannian nature of the submersion. For example, if we change the metric on each vertical space V_p in such a way that the volume form of M restricted to V_p is unchanged, then, by Remark 3.6, $\phi : M \to N$ becomes a Riemannian submersion with minimal fibres. Equally well, we may change the metric of M on each horizontal space in an arbitrary fashion. This does not affect total geodesity or minimality of the fibres. For example if we change the metric conformally on each horizontal space we obtain from the original Riemannian submersion with totally geodesic fibres a conformal submersion with totally geodesic (or minimal) fibres. This works in all dimensions, in particular if $\dim N = 2$ the maps obtained are all harmonic morphisms.

(iii) Define $\phi : S^3 \times S^1 \to S^2$ as the composition of the natural projection $S^3 \times S^1 \to S^3$ and the Hopf map $S^3 \to S^2$. Then with standard metrics, ϕ is a Riemannian submersion with totally geodesic fibres (which are tori). Altering the metric on $S^3 \times S^1$ as above gives examples of harmonic morphisms $M^4 \to N^2$ which are Riemannian or conformal submersions with totally geodesic or minimal fibres.

(iv) Given any smooth locally trivial fibre space $F \to M \overset{\phi}{\to} N$ with all manifolds connected and compact fibre F, by Rummler (see Kamber and Tondeur [1]) there is a metric on M such that the foliation defined by ϕ is minimal and Riemannian.

4. SECTIONS OF THE TWISTOR BUNDLES

Let V be a 2-dimensional distribution in a 4-dimensional Riemannian manifold M^4 and let H be the associated horizontal 2-dimensional distribution. From now on we shall assume that M^4 is oriented. We may locally choose orientations for each V_p and H_p so that the combined orientation of $V_p \oplus H_p = T_pM$ is that of M^4, we then define almost complex structures J^V, J^H on each V_p and H_p as in §3 by rotation through $+\frac{\pi}{2}$. Note that changing the orientation of V_p changes that of H_p also and replaces (J^V, J^H) by $(-J^V, -J^H)$, all results below will be independent of this change, therefore, there is no loss of generality in assuming J^V and J^H are globally chosen. The Gauss section of V, $\gamma : M \to G_2^0(M^4)$, then maps into the Grassmanian bundle $G_2^0(TM^4)$ of <u>oriented</u> planes. The almost complex structures J^V and J^H may be combined to give almost complex structures $J^1 = (J^V, J^H)$ and $J^2 = (-J^V, J^H)$ on each T_pM^4, note that J^1 is orientation preserving but J^2 is orientation reversing. Let Z^+ (resp. Z^-) be the fibre bundle over M whose fibre at p is all metric almost complex structures on T_pM which are orientation preserving (resp. reversing); these are the well-known <u>twistor bundles of M</u> (see Eells-Salamon [1]). The distribution V defines sections $\gamma^1 : M \to Z^+$, $\gamma^2 : M \to Z^-$ by $\gamma^1(p) = J^1$, $\gamma^2(p) = J^2$ (where J^1, J^2 both act on T_pM). We shall establish holomorphicity properties of these for the sort of distribution which arises from a harmonic morphism.

Note that if M is an open subset of Euclidean space $\mathbb{R}^m$ the twistor bundles of M are trivial $Z^{\pm} = M \times S^2$ and there is a well-known holomorphic bijection $G_2^0(\mathbb{R}^4) \simeq S^2 \times S^2$. To generalize this to the bundle case we define for each $p \in M$, a map $j_p : G_2^0(T_pM^4) \to Z_p^+ \times Z_p^-$ by $V_p \to (J^1, J^2)$.

This is clearly bijective with inverse given by $(J^1, J^2) \to V_p = \{X \in T_pM : J^1X = -J^2X\}$. Let $\mathcal{L}_{J^1}(V_p, H_p) \subset \mathcal{L}(V_p, H_p) \simeq T^f_{V_p} G^0_2(TM^4)$ be the subspace of all maps $A \in \mathcal{L}(V_p, H_p)$ which commute with J^1, define $\mathcal{L}_{J^2}(V_p, H_p)$ similarly. Then the differential of i

$$di_p : T_{J^1}Z^+_p \times T_{J^2}Z^-_p \to T_{V_p} G^0_2(T_pM^4)$$

clearly maps $T_{J^1}Z^+_p \times 0$ onto $\mathcal{L}_{J^2}(V_p, H_p)$ and $0 \times T_{J^2}Z^-_p$ onto $\mathcal{L}_{J^1}(V_p, H_p)$ thus we may use di_p to identify the tangent spaces to the fibre $T^f_{J^1}Z^+ = T_{J^1}(Z^+_p)$ and $T^f_{J^2}Z^+ = T_{J^2}(Z^+_p)$ with $\mathcal{L}_{J^2}(V_p, H_p)$ and $\mathcal{L}_{J^1}(V_p, H_p)$ respectively.

Since $di_p : \mathcal{L}_{J^2}(V_p, H_p) \times \mathcal{L}_{J^1}(V_p, H_p) \to \mathcal{L}_J(V_p, H_p)$ is then given by $(B,C) \to B + C$, its inverse is given by

$$dj_p : \mathcal{L}_J(V_p, H_p) \to \mathcal{L}_{J^2}(V_p, H_p) \times \mathcal{L}_{J^1}(V_p, H_p)$$

$$A \to (\tfrac{1}{2}(A - J^2AJ^2),\ \tfrac{1}{2}(A - J^1AJ^1)).$$

Give $T^f_{J^1}Z^+ = \mathcal{L}_{J^2}(V_p, H_p)$ the almost complex structure

$$B \to -B \circ J^V = J^H \circ B$$

and $T^f_{J^2}Z^+ = \mathcal{L}_{J^1}(V_p, H_p)$ the almost complex structure

$$C \to -C \circ J^V = -J^H \circ C.$$

These are chosen so that, for each p

$$j_p : G^0_2(T_pM) \to Z^+_p \times Z^-_p$$

is biholomorphic with respect to the complex structure J^f on $G^0_2(T_pM)$ (see §3, equation 1) but with respect to the complex structure $\tilde{J}^f$

(equation 2 of §3) the first component of j_p is holomorphic but the second is antiholomorphic.

Theorem 4.1 Let V be a 2-dimensional distribution on a 4-dimensional Riemannian manifold. Then V is integrable minimal and conformal if and only if the section

$$\gamma^1 : M^4 \to Z^+$$

is holomorphic with respect to the almost complex structure J^2 on M^4 and the section

$$\gamma^2 : M^4 \to Z^-$$

is antiholomorphic with respect to the almost complex structure J^1 on M^4.

Proof That $\gamma : M \to G_2^0(TM)$ is vertically antiholomorphic with respect to J^f and horizontally holomorphic with respect to $\tilde{J}^f$ translate under the isomorphism $T^f G_2^0(TM) = T^f_{J_1} Z^+ \times T^f_{J_2} Z^-$ into the given statement.

Corollary 4.2 Let $\phi : M^4 \to N^2$ be a submersive harmonic morphism from a four-dimensional Riemannian manifold to a Riemann surface. Then the Gauss section $\gamma^1 : M^4 \to Z^+$ is holomorphic w.r.t. J^2 and $\gamma^2 : M^4 \to Z^-$ is anti-holomorphic w.r.t. J^1.

Example 4.3 Let (M^4,g,J) be a Kähler surface and let $\phi : M^4 \to N^2$ be a holomorphic submersion into a Riemann surface. Then the resulting foliation has leaves which are complex submanifolds of M^4. With the correct orientation on the vertical and horizontal spaces we see that $J^1 = J$. Thus $\gamma^1 : M^4 \to Z^+$ is covariantly constant and $\gamma^2 : M^4 \to Z^-$ is holomorphic (with respect to J). Conversely, let (M^4,g) be any oriented

4-dimensional Riemannian manifold and V a 2-dimensional oriented distribution on M. Then if γ^1 is covariantly constant on M^4, the almost Hermitian manifold (M^4,g,J^1) is Kähler and V is an almost complex distribution on (M^4,J^1). If V is integrable this means that the leaves of V are complex submanifolds of (M^4,J^1).

Application to classification of harmonic morphisms Let M^m be a fixed Riemannian manifold and let k be a fixed integer, $0 < k < m$. Let $\mathcal{F}_k(M)$ denote the set of all k-dimensional foliations which are associated to a submersion from M onto some smooth manifold. Say two submersions are equivalent if and only if they have the same fibres; thus we have a bijection:

$$\left\{\begin{array}{l}\text{equivalence classes of submersions from } M\\ \text{to some } (m-k)\text{-dimensional manifold.}\end{array}\right\} \longleftrightarrow \mathcal{F}_k(M)$$

In the case $m = 4$, $k = 2$, we also have a bijection:

$$\mathcal{F}_2(M^4) \longleftrightarrow \text{smooth sections of } Z^+ \times Z^-.$$

Theorem 4.4 Let M^4 be a 4-dimensional Riemannian manifold. Then the above bijections restrict to bijections:

$$\left\{\begin{array}{l}\text{equivalence classes of submersive harmonic}\\ \text{morphisms from } M^4 \text{ to some Riemann surface}\end{array}\right\} \longleftrightarrow \left\{\begin{array}{l}\text{2-dimensional conformal}\\ \text{minimal foliations in } \mathcal{F}_2(M^4)\end{array}\right\}$$

$$\longleftrightarrow \left\{\begin{array}{l}\text{smooth sections } (\gamma^1,\gamma^2) \text{ of } Z^+ \times Z^- \text{ with}\\ \gamma^1\ J^2\text{-holomorphic and } \gamma^2\ J^1\text{-antiholomorphic.}\end{array}\right\}$$

Proof Note firstly that the first map (harmonic morphisms $\to$ foliations) is surjective because given any conformal minimal foliation in $\mathcal{F}_2(M^4)$, there is, by definition, a submersion $\phi : M^4 \to N^2$ to which it is associated. This submersion has minimal fibres and by conformal invariance of the horizontal metric in vertical directions (see the development after Remark 3.4) the horizontal metric clearly descends to a conformal structure on N^2 such that ϕ is horizontally conformal. Thus the first map is a bijection. That the second map (foliations $\to$ sections of $Z^+ \times Z^-$) is a bijection is Theorem 4.1.

Remarks (i) Two submersive harmonic morphisms $\phi : M \to N$, $\phi' : M \to N'$ with connected fibres are equivalent if and only if $\phi' = \psi \circ \phi$ for some conformal diffeomorphism $\psi : N \to N'$.

(ii) The Proposition is analogous to the Classification Theorem for conformal minimal immersions (or harmonic maps) of Eells and Salamon [1,2].

5. HARMONICITY OF THE GAUSS SECTION

Let V be a k-dimensional distribution on a Riemannian manifold M and let $\gamma : M \to G_k(TM)$ be its Gauss section. We wish to discuss harmonicity properties of γ. The section γ is called (a) harmonic (section) if its fibre tension field

$$\tau^f(\gamma) = \text{Trace } \nabla^f(d\gamma^f)$$

vanishes. Here $d\gamma^f : TM \to T^fG_k(TM)$ is the fibre component of the differential (see §3) and ∇^f denotes the connection on $T^fG_k(TM)$ or on $T^*M \otimes T^fG_k(TM)$ induced in a natural way from that on M (see C.M. Wood [1] for more details).

Remark If M is an open subset of Euclidean space $\mathbb{R}^m$ the tangent bundle is trivial and so we have the Gauss map $\gamma : M \xrightarrow[\text{section}]{\text{Gauss}} G_k(TM) = M \times G_k(\mathbb{R}^m) \longrightarrow G_k(\mathbb{R}^m)$ where the last map is the canonical projection. Then the Gauss section is a harmonic section if and only if the Gauss map is a harmonic map and all results below apply to this special case.

To study the trace we decompose it according to the decomposition $TM = V \oplus H$:

$$\text{Trace } \nabla^f(d\gamma^f) = \text{Trace } \nabla^f(d\gamma^f)|_{V \times V} + \text{Trace } \nabla^f(d\gamma^f)_{H \times H} ;$$

in terms of an orthonormal frame $e_1, \ldots, e_m$ for TM with $e_1, \ldots, e_{m-k}$ a frame for H and $e_{m-k+1}, \ldots, e_m$ a frame for V:

$$\sum_{j=1}^{m} \nabla^f d\gamma^f(e_j, e_j) = \sum_{i=1}^{m-k} \nabla^f d\gamma^f(e_i, e_i) + \sum_{\alpha=m-k+1}^{m} \nabla^f d\gamma^f(e_\alpha, e_\alpha).$$

We thus study $\text{Trace } \nabla^f(d\gamma^f)|_{V \times V}$ for a distribution V with various properties. Note that, as in §3, we can regard $d\gamma^f$ as having values in $\mathcal{L}(V,H)$ via the isomorphism $\mathcal{L}(V,H) = T^fG_k(TM)$. This isomorphism is connection preserving (C.M. Wood [1]), and ∇^f will thus also denote the connection on $\mathcal{L}(V,H)$ or the induced connection on $T^*M \otimes \mathcal{L}(V,H) = \mathcal{L}(TM \otimes V,H)$.

<u>Proposition 5.1</u> Let V be an integrable distribution. Then

$$\text{Trace } \nabla^f(d\gamma^f)|_{V\times V} = \overset{\circ}{\nabla}(k\mu_V) - \mathcal{H}\circ \text{Ric}_V$$

i.e. $\{\text{Trace } \nabla^f(d\gamma^f)|_{V\times V}\}(W) = \overset{\circ}{\nabla}_W(k\mu_V) - \mathcal{H}(\text{Ric}_V(W))$ for any $p \in M$, $W \in V_p$. Here k denotes the dimension of V and μ_V denotes the mean curvature of the distribution (see §3), thus $k\mu_V$ = trace of the second fundamental form of V; $\overset{\circ}{\nabla}$ denotes the Bott partial connection, i.e.,

$$\overset{\circ}{\nabla}_W(k\mu_V) = \mathcal{H}[W,k\mu_V] = \mathcal{H}(\nabla^M_W k\mu_V - \nabla^M_{k\mu_V} W) \quad \text{and}$$

$$\text{Ric}_V(W) = \sum_{\alpha=m-k+1}^{m} R(W,e_\alpha)e_\alpha \quad \text{where } \{e_\alpha\} \text{ is an orthonormal basis for } V_p.$$

Further the term $\mathcal{H}\circ \text{Ric}_V$ vanishes if M is a space form.

To prove the Proposition we need some Lemmas describing the amount of symmetry of $d\gamma^f$ and $\nabla^f d\gamma^f$.

<u>Lemma 5.2</u> If V is integrable,

$$d\gamma^f(Y)(Z) = d\gamma^f(Z)(Y)$$

for all $p \in M$, $Y,Z \in V_p$.

<u>Proof</u> $d\gamma^f(Y)(Z) - d\gamma^f(Z)(Y) = \mathcal{H}(\nabla^M_Y Z - \nabla^M_Z Y) = \mathcal{H}[Y,Z]$.

<u>Lemma 5.3</u> If V is integrable, for any $p \in M$, $X \in T_pM, Y,Z \in V_p$,

$$(\nabla^f d\gamma^f)(X,Y)(Z) - \nabla^f d\gamma^f(X,Z)(Y) = d\gamma^f(\mathcal{H}(\nabla^M_X Z))(Y) - d\gamma^f(\mathcal{H}(\nabla^M_X Y))(Z).$$

<u>Proof</u> $(\nabla^f d\gamma^f)(X,Y)(Z) = \nabla^f_X(d\gamma^f(Y)(Z)) - d\gamma^f(\nabla^M_X Y)(Z) - d\gamma^f(Y)(\nabla^V_X Z)$.

Here ∇^V denotes the connection on V induced from that of TM i.e.

$\nabla^V_X Z = \mathcal{V}(\nabla_X Z)$ where $\mathcal{V}: TM \to V$ is orthogonal projection. So

$$(\nabla^f d\gamma)(X,Y)(Z) = \nabla^f_X(d\gamma^f(Y)(Z))) - d\gamma^f(\nabla^V_X Y)(Z) - d\gamma^f(\mathcal{H}(\nabla^M_X Y))(Z) - d\gamma^f(Y)(\nabla^V_X Z)$$

$$= \nabla^f_X(d\gamma^f(Z)(Y)) - d\gamma^f(Z)(\nabla^V_X Y) - d\gamma^f(\mathcal{H}(\nabla^M_X Y)(Z) - d\gamma^f(\nabla^V_X Z)(Y)$$

(using Lemma 5.2)

$$= \nabla^f_X(d\gamma^f(Z)(Y)) - d\gamma^f(\nabla^M_X Z)(Y) - d\gamma^f(Z)(\nabla^V_X Y) - d\gamma^f(\mathcal{H}(\nabla^M_X Y))(Z)$$

$$+ d\gamma^f(\mathcal{H}(\nabla_X Z))(Y)$$

$$= \nabla^f d\gamma^f(X,Z)(Y) - d\gamma^f(\mathcal{H}(\nabla^M_X Y))(Z)) + d\gamma^f(\mathcal{H}(\nabla^M_X Z))(Y).$$

The Lemma follows.

Lemma 5.4 Let V be an arbitrary distribution (not necessarily integrable). Then for any $p \in M$, $X,Y \in T_pM$,

$$\nabla^f d\gamma^f(X,Y) - \nabla^f d\gamma^f(Y,X) = \mathcal{H} \circ R(X,Y)|_V$$

i.e. for any $Z \in V_p$,

$$\nabla^f d\gamma^f(X,Y)(Z) - \nabla^f d\gamma^f(Y,X)(Z) = \mathcal{H}(R(X,Y)Z).$$

Here $R(X,Y): T_pM \to T_pM$ is the Riemann curvature operator.

In particular, if M is a space form, $\nabla^f d\gamma^f$ is symmetric.

Proof

$$\nabla^f d\gamma^f(X,Y)(Z) = (\nabla^f_X d\gamma^f(Y))(Z) - d\gamma^f(\nabla^M_X Y)(Z)$$

$$= \nabla^H_X(d\gamma^f(Y)(Z)) - d\gamma^f(Y)(\nabla^V_X Z) - d\gamma^f(\nabla^M_X Y)(Z)$$

$$= \mathcal{H}\{\nabla^M_X \mathcal{H}(\nabla_Y Z) - \nabla^M_Y \nabla^V_X Z - \nabla^M_{\nabla^M_X Y} Z\}$$

$$= \mathcal{H}\{\nabla^M_X \nabla^M_Y Z - \nabla^M_{\nabla^M_X Y} Z - \nabla^M_X \nabla^V_Y Z - \nabla^M_Y \nabla^V_X Z\}.$$

Interchanging X and Y and subtracting yields the Lemma.

Proof of Proposition 5.1 Let $\{e_\alpha\}$ be an orthonormal frame for V. Then

$$(\text{Trace } \nabla^f d\gamma^f|_{V\times V})(Y) = \nabla^f d\gamma^f(e_\alpha, e_\alpha)(Y) \quad \text{(we adopt the summation convention)}$$

$$= \nabla^f d\gamma^f(e_\alpha, Y)(e_\alpha) + d\gamma^f(\mathcal{H}(\nabla^M_{e_\alpha} Y))(e_\alpha) - d\gamma^f(\mathcal{H}(\nabla^M_{e_\alpha} e_\alpha))(Y)$$

(using Lemma 5.3)

$$= \nabla^f d\gamma^f(Y, e_\alpha)(e_\alpha) + \mathcal{H}(R(e_\alpha, Y)(e_\alpha))$$

$$+ d\gamma^f(\mathcal{H}(\nabla^M_{e_\alpha} Y))(e_\alpha) - d\gamma^f(\mathcal{H}(\nabla^M_{e_\alpha} e_\alpha))(Y).$$

Now $\nabla^f d\gamma^f(Y, e_\alpha)(e_\alpha) = \nabla^H_Y(d\gamma^f(e_\alpha)(e_\alpha)) - d\gamma^f(\nabla^M_Y e_\alpha)(e_\alpha) - d\gamma^f(e_\alpha)(\nabla^V_Y e_\alpha)$.

Now decomposing $\nabla^M_Y e_\alpha = \mathcal{H}(\nabla_Y e_\alpha) + \nabla^V_Y e_\alpha$ note that the sum

$$d\gamma^f(\nabla^V_Y e_\alpha)(e_\alpha) = \langle \nabla^V_Y e_\alpha, e_\beta \rangle d\gamma^f(e_\alpha)(e_\beta)$$

is zero since $d\gamma^f$ is symmetric whereas the inner product is antisymmetric in α, β. Similarly $d\gamma^f(e_\alpha)(\nabla^V_Y e_\alpha) = 0$.

Finally $\mathcal{H}(\nabla^M_{e_\alpha} Y) = \mathcal{H}(\nabla^M_Y e_\alpha)$ since V is integrable, hence,

$$(\text{Trace } \nabla^f d\gamma^f|_{V\times V})(Y) = \nabla^H_Y(k\mu_V) - d\gamma^f(\mathcal{H}(\nabla_Y e_\alpha))(e_\alpha) - \mathcal{H}(\text{Ric}_V(Y))$$

$$+ d\gamma^f(\mathcal{H}(\nabla_Y e_\alpha))(e_\alpha) - d\gamma^f(k\mu_V)(Y).$$

On interpreting $d\gamma^f$, this gives desired formula.

Theorem 5.5 Let V be a Riemannian foliation with integrable horizontal distribution on a space form. Then the Gauss section is harmonic if and only if the leaves of V have parallel mean curvature i.e. $\nabla^H_Y(\mu_V) = 0$ for all $p \in M$, $Y \in V_p$.

Proof

$$\text{Trace } \nabla^f d\gamma^f|_{V\times V} = \overset{\circ}{\nabla}{}^H_Y(\mu_V) \quad \text{whereas}$$

$$\text{Trace } \nabla^f d\gamma^f|_{H\times H} = 0 \quad \text{(see remark).}$$

Since $\tau^f(\gamma)$ is the sum of these the Theorem follows.

Remark 5.6 (i) In the case of a codimension one foliation, the horizontal distribution is automatically integrable and this result was obtained by C.M. Wood [1].

(ii) Proposition 5.1 applied to H actually tells us about $\gamma^{\perp}$ rather than γ viz:

$$\text{Trace } \nabla^{f}(d\gamma^{\perp})^{f}|_{H \times H} = 0.$$

But clearly taking the adjoint of a mapping gives a connection preserving isomorphism

$$\mathcal{L}(V,H) \simeq \mathcal{L}(H,V)$$

under which $d\gamma(Z)$ corresponds to $-d\gamma^{\perp}(Z)$ for all $Z \in TM$ and so it follows that $\text{Trace } \nabla^{f}d\gamma^{f}|_{H \times H} = 0$.

Proposition 5.1 required integrability of V to give us symmetry properties such as Lemma 5.2 without which we do not get manageable formulae. However there is one situation in which we can handle non-integrability: Recall that a distribution V is called totally geodesic if its second fundamental from $\zeta(Y,Z) = \frac{1}{2}\mathcal{H}(\overset{M}{\nabla}_{Y}Z + \overset{M}{\nabla}_{Z}Y)$ vanishes for all $p \in M$.

Proposition 5.7 Let V be a totally geodesic distribution not necessarily integrable. Then for any $p \in M$, $Z \in V_p$

$$(\text{Trace } \nabla^{f}d\gamma^{f}|_{V \times V})(Z) = L_V(Z) + \mathcal{H}(\text{Ric}_V(Z))$$

where $L_V(Z) = \sum \mathcal{H}(\overset{M}{\nabla}_{[Z,e_\alpha]}e_\alpha)$ (summing over an orthonormal basis $\{e_\alpha\}$ for V). (To calculate this quantity we must extend Z and the $\{e_\alpha\}$ to local sections of V but it only depends on the value of Z at p.)

To prove this we replace Lemmas 5.2 and 5.3 by results concerning the antisymmetry of $d\gamma^{f}$ and $\nabla^{f}d\gamma^{f}$:

Lemma 5.8 If V is a totally geodesic distribution on M

$$d\gamma^f(Y)(Z) + d\gamma^f(Z)(Y) = 0$$

for all $p \in M$, $Y,Z \in V_p$.

Proof This is just expressing the vanishing of the second fundamental form.

Lemma 5.9 If V is a totally geodesic distribution on M,

$$\nabla^f d\gamma^f(X,X)(Z) + \nabla^f d\gamma^f(X,Z)(X) = - d\gamma^f(\mathcal{H}(\overset{M}{\nabla}_X Z))(X)$$

for all $p \in M$, $X,Z \in V_p$.

Proof Extend X and Z to local sections of V.

Then

$$\nabla^f d\gamma^f(X,X)(Z) = \overset{H}{\nabla}_X(d\gamma^f(X)(Z)) - d\gamma^f(\overset{M}{\nabla}_X X)(Z) - d\gamma^f(X)(\overset{V}{\nabla}_X Z).$$

Decomposing $\overset{M}{\nabla}_X X = \overset{V}{\nabla}_X X + \mathcal{H}(\overset{M}{\nabla}_X X)$ we note that $\mathcal{H}(\overset{M}{\nabla}_X X) = 0$ by total geodesicity. Similarly

$$\nabla^f d\gamma^f(X,Z)(X) = \overset{H}{\nabla}_X(d\gamma^f(Z)(X)) - d\gamma^f(\overset{M}{\nabla}_X Z)(X) - d\gamma^f(Z)(\overset{V}{\nabla}_X X).$$

Adding and using Lemma 5.8 yields the result.

Proof of Proposition 5.7 We shall show that

$$\nabla^f d\gamma^f(X,X)(Z) = d\gamma^f(\mathcal{H}[Z,X])(X) - \mathcal{H}(R(X,Z)X)$$

for all $p \in M$ and $X,Z \in V_p$. Let $Z \in V_p$; by Lemmas 5.9 and 5.4,

$$\begin{aligned}
\nabla^f d\gamma^f(X,X)(Z) &= - \nabla^f d\gamma^f(X,Z)(X) - d\gamma^f(\mathcal{H}(\overset{M}{\nabla}_X Z)(X) \\
&= - \nabla^f d\gamma^f(Z,X)(X) - \mathcal{H} \circ R(X,Z)(X) - d\gamma^f(\mathcal{H}(\overset{M}{\nabla}_X Z)(X) \\
&= - \overset{H}{\nabla}_Z(d\gamma^f(X)(X)) + d\gamma^f(\overset{M}{\nabla}_Z X)(X) \\
&\quad + d\gamma^f(X)(\overset{V}{\nabla}_Z X) - \mathcal{H}(R(X,Z)(X)) - d\gamma^f(\mathcal{H}(\overset{M}{\nabla}_X Z))(X) \\
&= - \mathcal{H}(R(X,Z)(X)) + d\gamma^f(\mathcal{H}[Z,X])(X)
\end{aligned}$$

using Lemma 5.8.

Setting $X = e_\alpha$, where e_α is an orthonormal basis for V, yields the result.

Remark 5.10 Note how Proposition 5.7 reduces to Proposition 5.1 in the integrable case.

Conformal foliations of codimension 2 We now give harmonicity conditions for the Gauss section of a conformal foliation of codimension two - such foliations arise, for example, from horizontally conformal submersions of a Riemannian manifold onto a surface. Recall that a foliation is conformal if and only if the horizontal distribution is umbilic. We thus study umbilic distributions. We shall use the holomorphicity properties established in §3.

Proposition 5.11 Let V be a two-dimensional umbilic distribution on a Riemannian manifold M. Then

$$(\text{Trace}\ \nabla^f d\gamma^f)_{V \times V} = -\, d\gamma^f(2\mu_V) + d\gamma^f(\mathcal{H}[X,Y]) \circ J^V + \mathcal{H} \circ R(X,Y) \circ J^V .$$

Here for any $p \in M$ we orient V_p arbitrarily and then let X and Y be an oriented orthonormal basis; J^V is rotation through $+\frac{\pi}{2}$. (We see that the right-hand side is independent of the orientation chosen.)

Remark 5.12 This formula reduces to that in Proposition 5.1 in the case of a totally geodesic distribution $(\mu_V = 0)$.

To prove this we need some lemmas relating various connections and almost complex structures. Suppose that we have chosen an orientation on V, as before this can always be done locally and our results will be independent of the actual choice.

Lemma 5.13 Let J^f be the almost complex structure defined on $\mathcal{L}(V_p,H_p)$ for each $p \in M$ by $J^f(A) = -J^V \circ A$ (as in §2). Then $\nabla^f J = 0$ i.e. $\nabla^f_Z(J(A)) = J(\nabla^f_Z A)$ for all sections Z of TM and A of $\mathcal{L}(V,H)$.

Proof For any $p \in M$, $Z \in T_pM$, $A \in \mathcal{L}(V_p,H_p)$, extending A to a local section of $\mathcal{L}(V,H)$ we have

$$\begin{aligned}(\nabla^f_Z J)(A) &= (\nabla^f_Z(J(A)) - J(\nabla^f_Z A) \\ &= \nabla^f_Z(-J^V \circ A) + J^V \circ \nabla^f_Z A \\ &= 0 \text{ by "product" rule and Lemma 2.5.}\end{aligned}$$

As in §3, let V' and V'' denote the $+i$ and $-i$ eigenspaces of the complex linear extension of J to V^c.

It is now convenient to introduce complex notation: if X and Y is an orthonormal frame for V we write $Z = X - iY$, $\bar{Z} = X+iY$; these are local sections of V' and V'' respectively.

Lemma 5.14 For an arbitrary 2-dimensional distribution V,

$$\begin{aligned}\nabla^f d\gamma^f(\bar{Z},Z) - \nabla^f d\gamma^f(Z,\bar{Z}) &= \mathcal{H} \circ R(\bar{Z},Z) \\ &= -2i\ \mathcal{H} \circ R(X,Y).\end{aligned}$$

Proof Immediate from Lemma 5.4.

Proof of Proposition 5.11 We have

$$\begin{aligned}\text{Trace } \nabla^f d\gamma^f|_{V \times V} &= \tfrac{1}{2}\{\nabla^f d\gamma^f(\bar{Z},Z) + \nabla^f d\gamma^f(Z,\bar{Z})\} \\ &= \nabla^f d\gamma^f(Z,\bar{Z}) - i\mathcal{H} \circ R(X,Y).\end{aligned}$$

But
$$\begin{aligned}\nabla^f d\gamma^f(Z,\bar{Z}) &= \nabla^f_Z(d\gamma(\bar{Z}) - d\gamma(\nabla^M_Z \bar{Z}) \\ &= \nabla_Z(d\gamma(\bar{Z})) - d\gamma(\nabla^V_Z \bar{Z}) - d\gamma(\mathcal{H}(\nabla_Z \bar{Z})).\end{aligned}$$

Now by Proposition 3.3, by umbilicity $d\gamma^f$ is vertically holomorphic so that $d\gamma^f(\bar{Z})$ is of type (0,1). By Lemma 5.13 ∇^f_Z preserves types so $\nabla^f_Z d\gamma^f(\bar{Z})$ is also of type (0,1). Looking at the next term, by Lemma 2.5, $\nabla^V_Z \bar{Z}$ is of type (0,1) and by holomorphicity so is $d\gamma^f(\nabla^V_Z \bar{Z})$. Thus taking (1,0) parts we have

$$\begin{aligned}\{\nabla^f d\gamma^f(Z,\bar{Z})\}^{(1,0)} &= -d\gamma^f(\mathcal{H}(\nabla_Z \bar{Z}))^{(1,0)} \\ &= \{-d\gamma^f(2\mu_V) - i\,d\gamma^f(\mathcal{H}[X,Y])\}^{(1,0)}.\end{aligned}$$

Substituting, we get

$$\begin{aligned}\{\text{Trace } \nabla^f d\gamma^f|_{V\times V}\}^{(1,0)} &= \{-d\gamma^f(2\mu_V) - i\,d\gamma^f(\mathcal{H}[X,Y]) - i(\mathcal{H}\circ R(X,Y))\}^{(1,0)} \\ &= \{-d\gamma^f(2\mu_V) - J^f d\gamma^f(\mathcal{H}[X,Y]) - J^f(\mathcal{H}\circ R(X,Y))\}^{(1,0)}\end{aligned}$$

since $J^f = i$ on the (1,0) eigenspace.

Since $\text{Trace } \nabla^f d\gamma^f|_{V\times V}$ is real it is determined by its (1,0) part and the Proposition follows.

Remark 5.15 (i) For those who would prefer a proof using only real notation, this can be given by writing

$$\begin{aligned}\text{Trace } \nabla^f d\gamma^f|_{V\times V} &= \nabla^f d\gamma^f(X,X) + \nabla^f d\gamma^f(Y,Y) \\ &= \nabla^f d\gamma^f(X,J^V Y) + \nabla^f d\gamma^f(Y,J^V X).\end{aligned}$$

We then compute using Lemmas 5.13, 5.14.

(ii) For a conformal foliation, Proposition 5.11 applied to the horizontal distribution gives us $\text{Trace } \nabla^f(d\gamma^\perp)^f|_{H\times H}$.

By combining Proposition 5.11 applied to the horizontal distribution and Proposition 5.1 applied to the vertical distribution we can now study harmonicity of the Gauss section of a conformal minimal foliation:

Theorem 5.17 Let V be a conformal minimal foliation of dimension 2 on a space form of dimension 4. Then

(i) If the leaves are totally geodesic, then the Gauss section $\gamma : M \to G_2(TM)$ is harmonic.

(ii) If the horizontal distribution is integrable, γ is harmonic if and only if the foliation is Riemannian or the leaves are totally geodesic.

(iii) If the foliation is Riemannian then γ is harmonic if and only if the horizontal distribution is integrable or the leaves are totally geodesic.

<u>Proof</u> By Proposition 5.1, Trace $\nabla^f(d\gamma^f)|_{V \times V} = 0$. By Proposition 5.11 applied to $\gamma^\perp$ and H,

$$\text{(1)} \qquad \text{Trace}\, \nabla^f d\gamma^{\perp f}|_{H \times H} = d\gamma^{\perp f}(2\mu_H) + d\gamma^{\perp f}(\mathcal{V}[X,Y]) \circ J^H .$$

Here μ_H denotes the mean curvature vector of the horizontal distribution H and (X,Y) is an oriented orthonormal basis for H. Thus γ is harmonic if the right-hand side of (1) vanishes.

(i) If the leaves of the foliation V are totally geodesic, then $d\gamma^{\perp f}$ is zero on all vectors of V so that both terms of (1) vanish.

(ii) If the horizontal distribution is integrable, the last term of (1) vanishes. The first term on the right-hand side vanishes if and only if <u>either</u> $\mu_H = 0$, i.e. H is totally geodesic, <u>or</u> $d\gamma^{\perp f}$ is zero on the non-zero vertical vector μ_H; since γ is vertically antiholomorphic, this last condition is equivalent to $d\gamma^{\perp f}$ being zero on all vertical vectors, i.e. V being totally geodesic. Note by real analyticity, one of the conditions must hold on all of M.

(iii) This time, the first term on the right-hand side of (1) vanishes and we argue similarly.

<u>Remark 5.18</u> (i) For an arbitrary foliation V of a Riemannian manifold Kamber and Tondeur [1] consider the orthogonal projection $\pi : TM \to H$ as an H-valued 1-form. Then $d\pi = 0$ always and $d^*\pi = 0$ if and only if the

foliation is minimal in which case they call V a harmonic foliation. (Here d and d^* are defined using the connection on H induced from the Levi-Civita connection on M.) Note that π and the Gauss section γ are related at a point $p \in M$ by $\nabla\pi(X)(Y) = d\gamma^f(X)(Y)$ $\forall X \in T_pM$, $Y \in V_p$ and $d^*\pi = \sum_{i=1}^{k} d\gamma^f(e_i)(e_i)$, where $\{e_i\}$ is an orthonormal basis of V_p, = mean curvature of the fibre. Clearly a harmonic foliation does not necessarily have harmonic Gauss section.

(ii) For M^m an open subset of Euclidean space and $\phi : M^m \to N^n$ a submersion onto a Riemannian manifold the differential $d\gamma$ of the Gauss map of ϕ, $\gamma : M^m \to G_{m-n}(\mathbb{R}^m)$ can be considered as a 1-form on M with values in $V^* \otimes H$. Baird [2] considers harmonicity of the $V^* \otimes \phi^{-1}TN$ valued 1-form $d\phi.d\gamma$ given by $d\phi.d\gamma : TM \xrightarrow{d\gamma} V^* \otimes H \xrightarrow{id \times d\phi} V^* \otimes \phi^{-1}TN$. Note that $d\phi.d\gamma(X)(Y) = -\nabla d\phi\,(X,Y)$ for $p \in M$, $X \in T_pM$, $Y \in V_p$. Also note that $d^*(d\phi.d\gamma) = \mathrm{Tr}\,\nabla d\phi|_{H \times H}.d\gamma + d\phi.\tau(\gamma)$ so that for a submersion with $\mathrm{Tr}\,\nabla d\phi|_{H \times H} = 0$ - for example for a Riemannian submersion (and see §2) - $d^*(d\phi.d\gamma) = 0$ if and only if γ is harmonic. In general harmonicity of the 1-form $d\phi.d\gamma$ depends on the map ϕ whereas harmonicity of γ depends only on the associated foliation on M.

REFERENCES

P. BAIRD [1], Harmonic maps with symmetry, harmonic morphisms and deformations of metrics, Research Notes in Mathematics 87, Pitman (1983).

P. BAIRD [2], The Gauss map of a submersion, Preprint (1983), Centre for Mathematical Analysis, Australian National University, Canberra.

P. BAIRD [3], Harmonic morphisms onto Riemann surfaces and generalized analytic functions, Preprint (1985), University of Melbourne.

P. BAIRD and J. EELLS [1], A conservation law for harmonic maps, Geometry Symposium Utrecht 1980, Lecture Notes in Mathematics 894, Springer-Verlag (1981), 1-25.

S.-S. CHERN [1], Minimal surfaces in Euclidean space of N dimensions, Symposium in Honor of Marston Morse, Princeton University Press (1965), 187-198.

J. EELLS and S. SALAMON [1], Constructions twistorielles des applications harmoniques, C.R. Acad. Sci. Paris I, 296 (1983), 685-687.

J. EELLS and S. SALAMON [2], Twistorial construction of harmonic maps of surfaces into four-manifolds, Ann. Scuola Normale Pisa (to appear).

B. FUGLEDE [1], Harmonic morphisms between Riemannian manifolds, Ann. Inst. Fourier 28 (1978), 107-144.

R.E. GREENE and H.H. Wu [1], Embedding of open Riemannian manifolds by harmonic functions, Ann. Inst. Fourier 25 (1975), 215-235.

T. ISHIHARA [1], A mapping of Riemannian manifolds which preserves harmonic functions, J. Math. Kyoto Univ. 19 (1979), 169-174.

F.W. KAMBER and P. TONDEUR [1], Harmonic foliations, in : Harmonic maps, Proceedings, New Orleans 1980, Lecture Notes in Mathematics 949, Springer-Verlag (1982), 87-121.

ABCDEFGHIJ-AMS-89876